Mikhail A. Vorontsov
Walter B. Miller (Eds.)

Self-Organization in Optical Systems and Applications in Information Technology

Second Edition
With 105 Figures

Springer

Professor Mikhail A. Vorontsov

M.V. Lomonosov Moscow State University,
International Laser Center, Vorob'evy Gory,
Moscow 119 899, Russia
and
New Mexico State University,
Las Cruces, New Mexico 88003, USA

Dr. Walter B. Miller

U.S. Army Research Laboratory, Battlefield Environment Directorate,
White Sands Missile Range, New Mexico 88002, USA

Series Editor:

Professor Dr. Dr. h.c.mult. Hermann Haken

Institut für Theoretische Physik und Synergetik der Universität Stuttgart,
D-70550 Stuttgart, Germany
and
Center for Complex Systems, Florida Atlantic University,
Boca Raton, FL 33431, USA

The first edition appeared in the series as Volume 66.

Die Deutsche Bibliothek – CIP-Einheitsaufnahme
Self-organization in optical systems and applications in
information technology/Mikhail A. Vorontsov; Walter B. Miller
(ed.). – Berlin; Heidelberg; New York; Barcelona; Budapest; Hong
Kong; London; Milan; Paris; Santa Clara; Singapore; Tokyo:
Springer, 1998
  (Springer series in synergetics)
  ISBN 3-540-64125-4

ISSN 0172-7389

ISBN 3-540-64125-4 2nd Edition Springer-Verlag Berlin Heidelberg New York

ISBN 3-540-57086-1 Springer-Verlag Berlin Heidelberg New York

Typesetting: Camera-ready copy from the editors using a Springer T_EX macro package
SPIN 10669327   55/3144 - 5 4 3 2 1 0 - Printed on acid-free paper

This book is due for return not later than the
last date stamped below, unless recalled sooner.

Berlin
Heidelberg
New York
Barcelona
Budapest
Hong Kong
London
Milan
Paris
Santa Clara
Singapore
Tokyo

Springer Series in Synergetics

Editor: Hermann Haken

An ever increasing number of scientific disciplines deal with complex systems. These are systems that are composed of many parts which interact with one another in a more or less complicated manner. One of the most striking features of many such systems is their ability to spontaneously form spatial or temporal structures. A great variety of these structures are found, in both the inanimate and the living world. In the inanimate world of physics and chemistry, examples include the growth of crystals, coherent oscillations of laser light, and the spiral structures formed in fluids and chemical reactions. In biology we encounter the growth of plants and animals (morphogenesis) and the evolution of species. In medicine we observe, for instance, the electromagnetic activity of the brain with its pronounced spatio-temporal structures. Psychology deals with characteristic features of human behavior ranging from simple pattern recognition tasks to complex patterns of social behavior. Examples from sociology include the formation of public opinion and cooperation or competition between social groups.

In recent decades, it has become increasingly evident that all these seemingly quite different kinds of structure formation have a number of important features in common. The task of studying analogies as well as differences between structure formation in these different fields has proved to be an ambitious but highly rewarding endeavor. The Springer Series in Synergetics provides a forum for interdisciplinary research and discussions on this fascinating new scientific challenge. It deals with both experimental and theoretical aspects. The scientific community and the interested layman are becoming ever more conscious of concepts such as self-organization, instabilities, deterministic chaos, nonlinearity, dynamical systems, stochastic processes, and complexity. All of these concepts are facets of a field that tackles complex systems, namely synergetics. Students, research workers, university teachers, and interested laymen can find the details and latest developments in the Springer Series in Synergetics, which publishes textbooks, monographs and, occasionally, proceedings. As witnessed by the previously published volumes, this series has always been at the forefront of modern research in the above mentioned fields. It includes textbooks on all aspects of this rapidly growing field, books which provide a sound basis for the study of complex systems.

A selection of volumes in the Springer Series in Synergetics:

Preface

After the laser came into existence in 1960, basic experimental and theoretical research was focused on its behavior in the time domain. In this way, single and multimode operation and the effect of frequency locking as well as various kinds of spiking and ultrashort pulses were studied. Later, laser-light chaos was predicted and discovered. Sophisticated investigations concentrated on the study of the line width and of photon statistics. Though it was known from the very beginning that cavity modes may show different kinds of spatial patterns, in particular in the transverse direction, it was not until more recently that transverse spatial patterns caused by laser-light dynamics were discovered. Since laser action can be maintained only if the laser is continuously pumped from the outside, it is an open system. It is by now well known that there are a number of open systems in physics that may show spontaneous formation of spatial or temporal or spatio-temporal patterns. An important example is provided by fluid dynamics where a liquid layer heated from below may suddenly form hexagonal cells. In the center of each cell the fluid rises and sinks down at the borders. Also other patterns are observed such as rolls and stripe structures showing various kinds of defects. Similarly, certain chemical reactions can develop large-scale patterns such as concentric rings, spirals or, as was shown more recently, hexagonal and stripe patterns.

As I have shown a number of years ago, the occurence of such patterns, irrespective of the physical substratum, can be traced back to general principles of self-organization that are explored in the field of synergetics. More technically speaking, they can be explained by specific kinds of equations that I derived and called the generalized Ginzburg-Landau equations.

The discovery of spatial transverse patterns in lasers is of great importance in several ways. First of all, we discover that pattern formation is a wide-spread phenomenon in open systems. As will become more obvious from the articles in this book, lasers provide us with a wonderful means with which to study general phenomena of self-organization that are not so easily obtainable in chemical reactions or in fluid dynamics. One reason is the much shorter time scale in lasers, another one is the possibility to manipulate the individual modes in a sophisticated manner. Furthermore a number of technical applications become possible. As it seems, we are at the beginning of what may be called optical computers, including the optical synergetic computer.

This book, written by several pioneers of this new field, gives an excellent survey of the present state of the art and provides the reader with deep insights into the mechanisms by which patterns in lasers and in passive optical devices are formed.

I wish to congratulate Professor Mikhail Vorontsov, who himself has made a fundamental contribution to this field, for the excellent selection of authors and articles, and to both he and Dr. Miller for their editing work. I am sure that this book will be of great help to all scientists and engineers working on fundamental problems in optics or looking for important new applications. The book concludes with an excellent article by Professor Yuri Klimontovich on the general problem of self-organization.

Stuttgart, February 1995   H. Haken

To Professor Sergey Akhmanov

who had a profound impact on the development of nonlinear optics. Perhaps more profound was the impact of his personality on colleagues and students at Moscow State University.

Acknowledgments

Behind the publication of every book lies a separate and somewhat dramatic story. For our own particular case, we wish to take a moment and recognize several of the lead participants and acknowledge their roles. First and foremost, we thank the authors, without whose energy, dedication, and enthusiasm this book would not exist. Both Janet Vasiliadis and Boris Samson did a tremendous service to all by assisting as technical editors and providing moral support. Egor Degtiarev graciously prepared the chapter by S. A. Akhmanov and helped with preparation of the camera-ready manuscript. W. Firth's interest and support throughout the entire process was greatly appreciated. Last, but not least, we acknowledge the role played by all those at the Army Research Laboratory whose interest allowed the completion of this book up to the final act. In particular, we recognize the support of Don Veazey and Douglas Brown, and the ever-present help provided by Jennifer Ricklin. To all of these players we offer our sincerest thanks.

The Editors

Contents

CHAPTER 5

Transverse Traveling-Wave Patterns and Instabilities in Lasers
– Q. Feng, R. Indik, J. Lega, J.V. Moloney, A.C. Newell,

CHAPTER 6

Laser-Based Optical Associative Memories – M. Brambilla,
L.A. Lugiato, M.V. Pinna, F. Pratti, P. Pagani, and

CHAPTER 7

Pattern and Vortex Dynamics in Photorefractive Oscillators
– F.T. Arecchi, S. Boccaletti, G. Giacomelli, P.L. Ramazza,

CHAPTER 8

**From the Hamiltonian Mechanics to a Continuous Media.
Dissipative Structures. Criteria of Self-Organization –
Yu. L. Klimontovich** 217

Introduction
Self-Organization in Nonlinear Optics –
Kaleidoscope of Patterns

M. A. Vorontsov[1,2,3] *and W. B. Miller*[3]

[1] New Mexico State University, NM 88003, USA
[2] International Laser Center, Moscow State University, 119899 Moscow, Russia
[3] U.S. Army Research Laboratory, White Sands Missile Range, N.M. 88002, USA

Introductory Remarks

During all of its "coherent" lifetime optics, or rather should we say radio physics, the more mature and developed field for coherence and nonlinear problems, has been an assiduous discipline. Fourier optics, harmonics generation, parametric excitation, adaptive optics (the reader may continue this list), all these different directions taken by modern optics along with all of their underlying concepts appeared earlier in radio physics. It is easy to understand this friendship, with such strong and unequal rights, as the foundation for coherent and nonlinear optics was laid in the early 1960s by radio physicists unfaithful to their discipline.

Now we see new tendencies in optics; the former friendship is disappearing, and scientists who faithfully borrowed ideas from radio physics have begun to pay increasing attention to other disciplines. Hydrodynamics and the theory of nonequilibrium systems (or synergetics) are now the favorite subjects for imitation [NP, MY, Hak, CH]. It seems at last researchers have realized it is impossible to pay tribute to coherence all of one's life, and the multidimensional nature of an optical field is perhaps as important as its coherent nature. How long will this new attachment to synergetics and hydrodynamics last? Who knows? It certainly seems it won't be forever, but perhaps this new experience will at last help optics to reveal its own charming face. This book discusses these and related new trends.

1 What Is This Book About?

The first chapter by Akhmanov is an introduction to the subject of *information processing and nonlinear optics*. The information aspect of nonlinear optics is closely linked to the abilities of optical systems for different pattern generation and control. Information can be coded and processed as an optical pattern. From this point of view, the potential information capacity of nonlinear optical systems is determined by the number of different patterns (modes) that can exist and interact in a system. This means one of the critical factors of a nonlinear optical system – a potential candidate for information processing – is the *complexity* of its spatio-temporal dynamics.

Chapter 2 presents a discussion of how to design and control complex dynamics using rather simple nonlinear optical systems with two-dimensional feedback. The basic ideas of *artificial complexity* design in nonlinear 2-D optical systems derive in this chapter from classical synergetics and neural network theory. However, traditional approaches can be significantly expanded in optics.

How are simple optical patterns in nonlinear optical systems created? What parameters are responsible for pattern formation? How do defect structures appear? In Chapt. 3 Firth analyzes these problems using the presently most popular roll/hexagonal-type patterns in passive nonlinear systems with third-order nonlinearity (Kerr media).

In a Kerr-type medium the refractive index depends only on the intensity distribution, which gives rise to rather simple and convenient mathematical models for analysis. (Kerr-slice optical systems are very popular among theoreticians.) At the same time, for most materials the Kerr effect is too weak to be widely used in experiments. (Kerr-slice optical systems are very unpopular among experimentalists). The compromise solution suitable for both theoretical and experimental personalities is to change the type of nonlinearity. Rather complicated spatio-temporal dynamics appear in simple optical systems with two-level atomic media. The mathematical model for two-level atom optical systems is challenging enough to occupy several theoretical groups [Sil, LRT], but at the same time this type of nonlinearity is of no fear to experimentalists [SM, GVK]. Different instabilities in two-level atom devices are described in Chapt. 4.

Lasers with nonlinear active media are an excellent source of complexity [LOT, Lug, Wei, MN]. Transverse traveling waves, Eckhaus and zig-zag instabilities, optical turbulence – all these dynamic regimes in lasers cause *hydrodynamic* analogues in optics. In Chapt. 5 the mathematical models for both two-level and Raman lasers based on the Maxwell-Bloch equations are introduced. Analytical results and numerical simulation for traveling waves and different instabilities are also presented.

The complexity of a system's dynamic behavior by itself probably isn't sufficient to classify a system as an optical computer. An important problem is how to fit complexity into information processing? It is shown in Chapt. 6 that the coexistence of spatial modes in lasers, plus strong competition between these modes, allow us to consider the laser as an associative memory element or a specific kind of synergetic computer as suggested by Haken [FH]. A laser is capable of discriminating which of its spatial modes is most similar to an input image injected into the laser.

Complexity, competition (and certainly, money for research) are all needed, but what else is required for successful application of nonlinear optical systems to information processing? The controllability of the nonlinear dynamics is of great importance.

In Chapt. 7 by Arecchi and others photorefractive ring oscillator dynamics are considered. This system is of particular interest because dynamic behavior is represented by single and multimode regimes, intermode competition, periodic and chaotic alternation, and spatio-temporal chaos. By controlling the parameters of the system it is possible to transition between different dynamic regimes.

In the last chapter by Klimontovich self-organization processes are discussed from the point of view of macroscopic open system modern statistical theory. Results of this theory can be applied to different natural systems. This chapter underlines the importance of self-organization criteria that make it possible to determine the presence of order even in a turbulent pattern.

2 Nonlinear Optics: The Good Old Times

For a long time in nonlinear optics only problems of temporal dynamics were investigated. With this state of affairs, only longitudinal (along the optical axes) interactions were accounted for [Shn]. Removing transverse spatial effects was rather easy, at least in theory. It was enough to assume the optical field takes the form of a plane wave and that the propagation media and initial conditions are spatially uniform. Even in this artificial situation, when the primary mechanisms for manifestation of nonlinear effects are blocked, nonlinear system dynamics have a great variety. We observe practically all known dynamic regimes in radio physics, from trivial oscillations to complicated scenarios of chaos transition [OI].

Nevertheless, in some sense the limitations of these approaches were always obvious. What does pure temporal chaos of the spatially distributed optical field mean? If we prepare a nonlinear optical system having complicated temporal dynamics we cannot retain purity in spatial dimensionality. How do we prevent diffraction and diffusion in nonlinear media when both are always ready to destroy the ideal picture of pure temporal interactions?

The time has only now arrived to introduce the unused reserves of nonlinear wave dynamics – the laser beam's transverse spatial interactions. This first step is important primarily because by hiding in the tangle of nonlinear interactions in time and space, we deprive ourselves of the very pleasant opportunity to use the preferred method of factorizing spatial and temporal variables. Let us investigate this situation in more detail.

To describe the various aspects of problems concerning light/matter interactions, researchers now use rather complicated and cumbersome systems of equations. In this introduction, intended to attract readers to our subject, we describe the primary problems with simple mathematical models and reward the reader with a sample of the intriguing optical patterns that typically occur.

3 The First Model – Kerr-Slice/Feedback-Mirror System

The Kerr Slice/Feedback-Mirror model requires only two elements: a thin slice of a nonlinear medium and a feedback mirror (Fig. 1a) [Fir]. We need only two equations to describe the behavior of this system.

The first equation describes behavior of the light field's complex amplitude $A(\mathbf{r}, z, t)$

$$-2ik\frac{\partial A}{\partial z} = \nabla_\perp^2 A, \tag{1}$$

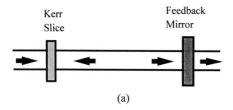

(a)

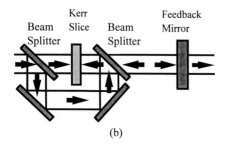

(b)

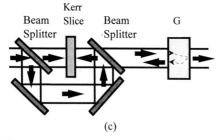

(c)

Fig. 1. Nonlinear 2-D feedback systems using a Kerr slice. (a) Kerr-slice feedback-mirror system, (b) nonlinear interferometer with 2-D feedback, (c) nonlinear interferometer with nonlocal interactions.

with boundary conditions determined outside the nonlinear slice

$$A(\mathbf{r},0,t) = I_0^{1/2} \exp\left[iu(\mathbf{r},t) + i\varphi(\mathbf{r})\right], \qquad (2)$$

where φ is phase distribution of the input beam.

The second equation is an example of the nonlinear matter's influence on the characteristics of the optical field. In our case of Kerr-type nonlinearity, this is the influence of matter on the phase $u(\mathbf{r},t)$ of the coherent wave passing through the slice

$$\tau\frac{\partial u}{\partial t} + u = D\nabla_\perp^2 u + K(I_0 + \mid A_{FB}(\mathbf{r},t)\mid^2). \qquad (3)$$

Here $\nabla_\perp^2$ is the Laplacian in the x and y directions describing transverse diffusion in the nonlinear medium with coefficient D, I_0 is the input field intensity, $\mid A_{FB}\mid^2 = \mid A(\mathbf{r}, z = L, t)\mid^2$ is feedback intensity (intensity inside the slice from the wave reflected from the feedback mirror), and L is the length of wave propagation.

We have several basic controlling parameters on which the dynamics depend: K (which includes wave number $k = 2\pi/\lambda$, nonlinearity coefficient n_2, and thickness of the slice l), the diffusion coefficient D, and the time response τ. All of these are characteristics of the medium. We also need three parameters that characterize diffraction: wave number k, diffractive length L, and laser beam aperture size a.

For convenience we decrease the number of parameters by combining them, so that $l_D = \sqrt{D}$ is the diffusive length and $F_a = a^2/(\lambda L)$ is the Fresnel parameter.

The Kerr-slice/feedback-mirror system is now commonly used as a classical model for nonlinear dynamics. Readers are referred to Chapt. 3 and articles [DF, VF] for a more detailed and rigorous treatment.[1] The photos in Fig. 4 of Chapt. 3 represent typical patterns from this system.

Using the Kerr-slice/feedback-mirror system as a model, we will demonstrate the basic manipulations characteristic of nonlinear optical dynamics.

4 Diffusion, Diffraction, and Spatial Scales

The traditional way to begin the theoretical study of nonlinear optical systems is to declare that the input field is a plane wave ($\varphi = 0$ and $F_a \gg 1$). This kind of declaration usually means we get to neglect diffraction. However, this is not quite true for systems capable of self-excitation. If we require that $F_a \gg 1$, we only estimate diffractive effects arising from the largest scale occurring in the problem, that is, the size of the laser beam aperture a. However, due to nonlinear interactions self-induced spatial phase inhomogeneities with size $l_h \ll a$ can arise in the system. We cannot ignore this possibility, for in doing so we would neglect the diffractive effects caused by such inhomogeneities (the scale l_h can be influenced by diffraction). This presents an interesting situation. Preparatory to ignoring diffraction we have to solve the complete problem, diffraction included. The only thing we can be certain of is that the spatial scale for self-induced inhomogeneities l_h should exceed the diffusion length l_D, which comprises the smallest scale of the problem (the diffusion cutoff). In this light, the second Fresnel parameter $F_{l_D} = l_D^2/(\lambda L)$ arises quite naturally. If L is so small that $F_{l_D} \gg 1$, we can neglect diffractive effects and instead of (1) write a trivial expression for the feedback field in the Kerr slice plane

$$A_{FB} = A(\mathbf{r}, L, t) = I_0^{1/2} \exp(iu + i\varphi_0). \tag{4}$$

Propagation of the wave causes only an additional spatially uniform phase shift φ_0.

As a result we have a spatially homogeneous intensity on the right side of equation (3), with only a spatially homogeneous solution of this equation, $u(\mathbf{r}, t) = 0$. Self-organization will not appear if diffraction is ignored.

---

[1] There is an optical scheme that is probably even simpler than the Kerr-slice/feedback-mirror system, which consists of only one nonlinear element and two coherent counter-propagating waves. Nevertheless, this simple optical scheme gives rise to complicated spatial-temporal dynamics (see Chapt. 4 and [CG, GB]).

Is it always true that only diffraction can produce patterns in nonlinear optics? Fortunately, the answer is no. We will show that using only diffusion-type interactions many different patterns can be produced. First it is necessary to change the Kerr-slice/feedback-mirror scheme slightly, as shown in Fig. 1b.

5 One More Scheme: The First Step Toward Optical Synergetics

This change gives us two coherent input beams with an interference pattern in the Kerr slice plane. We can now neglect diffraction without the danger of producing a trivial spatially uniform solution.

To correct our mathematical model all we need to do is correct the formula (4) for the feedback field by adding the reference wave with complex amplitude $I_0^{1/2} \exp(iu_0)$

$$A_{FB} = A(\mathbf{r}, L, t) + I_0^{1/2} \exp(iu_0). \tag{5}$$

Using (4), we obtain the expression for feedback intensity

$$I_{FB} = I_0[1 + \gamma \cos(u + \chi_0)], \tag{6}$$

where $\chi_0 = \varphi_0 - u_0$ is a spatially uniform phase and $\gamma \leq 1$ is the interference pattern visibility (another controlling parameter).

Substituting (6) into (3) gives us nonlinear diffusion or a Fisher-Kolmogorov-Petrovskii-Piskunov (FKPP)-type equation

$$\tau \frac{\partial u}{\partial t} + u = D\nabla_\perp^2 u + f(u), \tag{7}$$

where $f(u) = R[1 + \gamma \cos(u + \chi_0)]$ and R is a control parameter determined by the input field intensity I_0.

The FKPP-type equation is one of the cornerstones in the foundation of modern nonequilibrium systems theory [ML]. What is important here isn't that we have exactly a cosine-type nonlinearity, but that f is an "N"-type function. Perhaps this is the first thread to tie optics with chemistry and biology, from which the FKPP equation originated [Mur, KPP].

This equation is responsible for two effects: optical bistability (multistability) and switching waves [ML, FG, MZI]. The nonlinear interferometer with 2-D feedback shown in Fig. 1b is an optical model of a one-component reaction-diffusion nonequilibrium system (see also Chapt. 2).[2]

---

[2] Optical multistability, switching waves, and spatio-temporal oscillations were also observed in different interferometers using semiconductive nonlinear media [RRH, Ros].

6 Nonlocal Interactions; A Kaleidoscope of Patterns

Unlike chemistry and biology where the FKPP equation originated, in nonlinear optics we have many unique opportunities to design new types of spatial interactions. This can be accomplished by having a point in a laser beam cross section interact not only with its neighbors (local spatial interactions), but with distant points as well creating so-called "nonlocal" or "large-scale" interactions [VDP, VPS, Vor]. Figure 1c shows a typical system displaying these types of interactions. In fact, this is the same scheme as in the nonlinear interferometer with 2-D feedback, but instead of a mirror a special reflective optical element G is placed in the feedback loop. This element changes the direction of the passing rays. The ray trajectory depends on the transverse coordinate $\mathbf{r} = \{x, y\}$.[3] What occurs in the optical feedback is a form of mapping or coordinate transformation [AVL]. Excitation of the nonlinear medium at one point $\mathbf{r}$ causes a response at some distant point $\mathbf{r}^\star = G\mathbf{r}$ after the laser beam has passed through the feedback loop, where G is the coordinate transformation operator.

To implement the simplest types of coordinate transformation (field rotation along the optical axes, transverse shift, or scale change), it is possible to use rather simple optical elements, such as a Dove prism or a system of mirrors or lenses [Vor]. For more complicated spatial mapping computer-generated holograms can be used.

From a practical point of view, for actual creation of nonlinear optical systems with Kerr-type nonlinearity it is more convenient to use a liquid crystal light valve (LCLV) as a Kerr slice model (see [VKS, VoL] and Chapts. 2 and 3).

The basic schemes for this type of optical system with field transformation are shown in Fig. 2. This type of system is called the *Optical Kaleidoscope*, or in short form, the *OK-system*.[4] The system shown in Fig. 2b illustrates the OK-system, which is based on the polarization effect (the polarized interferometer). The principle of its operation is described in Chapt. 2 and [ALV].

The first experiments using an OK-type system were carried out at Moscow State University in 1988 [VDP, VIS, AhI]. During the course of these experiments different patterns were produced [Vor], including

- rolls and their defects
- 1-D and 2-D rotatory waves (see also [AkV, VIL, ILV, AVL])
- optical spirals
- concentric waves
- hexagonal-type patterns
- turbulent-type patterns
- patterns with complicated geometry
- evidence of the coexistence of different patterns.

---

[3] The simplest example of this optical element is a retroreflector, or a combination of lens and mirror.

[4] The OK-system is a nonlinear optical system using an LCLV with nonlocal interactions in the two-dimensional optical feedback.

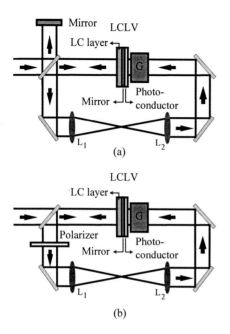

Fig. 2. Nonlinear optical systems with nonlocal interactions using a liquid crystal light valve (LCLV) – the OK-type systems. (a) Interferometric scheme using an LCLV in the regime of phase modulation. (The direction of the incident wave polarization is parallel to the optical axes of the liquid crystal molecules.) (b) Polarized interferometer. (The direction of incident wave polarization and the optical axes of the molecules of liquid crystal are at an angle $\pi/4$. The polarizer axes are orthogonal to incident-light polarization.)

In these first experiments with the OK-system the authors tried to exclude the influence of diffractive effects and investigate the process of pattern formation using pure diffusion-type interactions. Two methods were used. The first was to make the length of the feedback loop as short as possible, and the second was to use a pair of lenses placed into the optical feedback to form an image of the liquid crystal (LC) layer on the photoconductive layer at the back of the LCLV (see Fig. 2). Among recent experimental results using systems similar to the OK-system we can refer to [TNT, PRR, ALV, PRA].

Figure 3 gives examples of 2-D rotatory waves in the OK-system with field rotation. Attempts to exclude the influence of diffraction were not totally successful. Figure 3c shows an example of the coexistence of two different types of patterns.[5] The large white and black petals that totally occupy the laser beam aperture represent a typical 2-D rotatory wave caused by a pure diffusion-type interaction, while the small spots in the central part of the pattern are the result of a diffractive-type interaction – Akhseal-type patterns (discussed below).

---

[5] This pattern was obtained using the OK-system in collaboration with Firth's group.

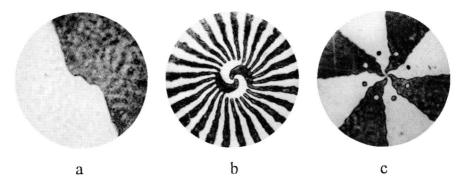

a   b   c

Fig. 3. Two-dimensional rotatory waves in the OK-system. (a) Two-petal rotating structure. (b) Rotating structure spatial bifurcation and the coexistence of two waves rotating in different directions. (c) Coexistence of two structures – rotatory wave and diffractive pattern (small spots in the center portion).

7 OK-Equation and "Dry Hydrodynamics"

Nonlinear dynamics of the OK-system with pure diffusion-type local interactions can be described through a slight correction in the nonlinear-diffusion equation (7). Simply replace vector $\mathbf{r}$ with $\mathbf{r}^\star$ in the nonlinear function $f(u(\mathbf{r}, t))$, and we obtain the OK-equation [Vor]:

$$\tau \frac{\partial u}{\partial t} + u = D\nabla_\perp^2 u + f(u(\mathbf{r}^\star, t)), \tag{8}$$

where $\mathbf{r}^\star = G\mathbf{r}$. This small change leads to a great variety of different solutions (patterns), as opposed to the single, lonely switching waves typical of the FKPP equation.

Considering nonlocal interactions leads us to an analogy. With just a slight transversal shift of the laser beam in the 2-D feedback at a distance $\mathbf{v} = \mathbf{r}^\star - \mathbf{r}$, where $|\mathbf{v}| \ll 1$, we can use the nonlinear term expansion in the OK-equation (8) and obtain an equation similar to that of rotating fluid dynamics [HLR, DDK]

$$\tau \frac{\partial u}{\partial t} + u = D\nabla_\perp^2 u + f(u(\mathbf{r}, t)) + (\mathbf{v} \cdot \nabla_\perp u) f'(u). \tag{9}$$

This is highly reminiscent of expressions responsible for turbulent regimes in hydrodynamics (*dry hydrodynamics*, see Chapt. 7 and [Are, Sta]). The resulting "artificial" turbulence offers many opportunities for controlling and measuring the primary characteristics of turbulence. Because it is possible to produce artificial turbulence in a thin layer of a nonlinear medium having a thickness $h \ll l_D$, we essentially have two-dimensional turbulence [DDK].[6] The decreasing of dimensionality is a challenge for producing new analytical results. But that isn't all we can do by removing diffraction.

---

[6] Typical parameters for an LCLV having an aperture $a = 20$ mm are $l = 10\mu$ and $l_D = 100\mu$.

8 One More Nonlinear Element: Two-Component Optical Reaction-Diffusion Systems

Simply add an additional nonlinear optical system similar to the OK-system as shown in Fig. 4 to create a two-component reaction-diffusion system (the sacred altar of classic synergetics) [Hkn, Kur, ML]:

$$\tau_u \frac{\partial u}{\partial t} = D_u \nabla_\perp^2 u + f(u, v, \mathbf{K}),$$
$$\tau_v \frac{\partial v}{\partial t} = D_v \nabla_\perp^2 v + g(u, v, \mathbf{K}). \tag{10}$$

In our case of two optical subsystems, the spatial variables $u(\mathbf{r}, t)$ and $v(\mathbf{r}, t)$ are functions that characterize the nonlinear phase modulation of each nonlinear element, τ_u and τ_v are relaxation times that determine the temporal scale of changes in the u and v variables, and f, g are nonlinear functions.

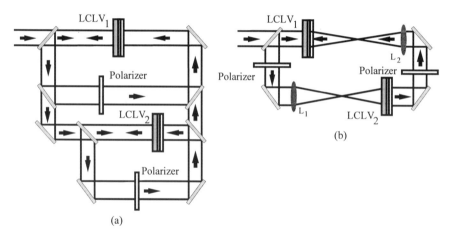

Fig. 4. Nonlinear optical systems using LCLVs to model two-component reaction-diffusion system dynamics. (a) System (polarized interferometer) with two reflective LCLVs and bounded feedback loops [DV]. (b) Optical scheme with reflective ($LCLV_1$) and transparent ($LCLV_2$) liquid crystal light valves.

The diffusion coefficients D_u and D_v set the scale of spatial interactions and the control parameters $\mathbf{K}$ determine the system's excitation level.

A system of equations like those in (10) provides a fairly simple mathematical model for analyzing a great variety of self-organization phenomena in diverse nonequilibrium systems. By altering the parameters, many different patterns typical of non-optical synergetics can be obtained, such as dissipative structures, traveling waves, and various autowave regimes (see [ML] and Chapt. 2).

In classical synergetics these dynamic regimes have been assembled from many disciplines, some from chemistry, others from plasma physics, and some from the arrhythmic palpitations of the heart or the movement of a protoplasm [VRC].

The possibility now exists for creating an optical *design kit* for studying different multicomponent reaction-diffusion systems. Combining one-component optical systems produces various multicomponent active systems having fantastically rich dynamics. Add to this the unique opportunity of changing and controlling the system's parameters (this isn't an experiment using a human heart), and watch this process in real time.

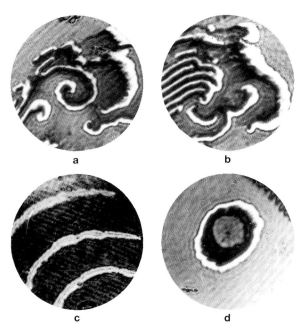

Fig. 5. Dynamic regimes in a two-component optical system [RVP, VRS]. (a) Periodically-induced spiral wave, (b) Coexistence of spiral wave and rolls, (c) Self-induced traveling pulses, (d) Leading center formed by external narrow laser beam in the central part of the aperture.

Figure 5 shows dynamic patterns typical of two-component systems, spiral waves (Fig. 5 a,b), traveling pulses (Fig. 5 c), and leading centers (Fig. 5 d). All these examples of spatio-temporal dynamics were obtained using an optical model of a two-component active system. The optical schematic for this system is shown in Fig. 4b (an additional transparent LCLV with $\tau_v \gg \tau_u$ was placed into the 2-D feedback loop [RVP, VRS, Vrt]).

By creating nonlocal interaction in two-component "reaction-diffusion" optical systems we can investigate unique dynamic regimes unknown in non-optical synergetics [DV].

9 Diffraction at Last; Rolls and Hexagons

Diffusion isn't the only mechanism that can provide local transverse interactions. With just a bit of diligence and effort, it is perhaps possible to avoid diffusion and design nonlinear optical systems having pure diffractive local transverse interactions.

Consider again the Kerr-slice/feedback-mirror system. In the previously discussed Kerr-slice/feedback-mirror model shown in Fig. 1a we used the plane-wave approximation. However, this isn't a pure diffusive approach, because the pattern-formation process in this system is due solely to diffraction caused by small inhomogeneities in the initial phase modulation $u(\mathbf{r}, 0)$ in (3). We can even try to ignore diffusion completely.

Suppose the initial phase modulation contains spatial spectral components (the periodic wave "rolls") with wavevectors $\mathbf{q}$,

$$\Phi_q = a_q \cos{(\mathbf{q} \cdot \mathbf{r} + \varphi_q)}.$$

Due to nonlinear interaction in the system, rolls with certain wavevectors grow while others are suppressed. The consequence of growth is that the phase modulation spatial components modulate the intensity in the Kerr-slice plane (positive feedback) after propagation over the distance L. Positive feedback will occur if the increasing intensity modulation gives rise to a corresponding increase in nonlinear phase modulation components (modes) with the same spatial frequency as $\mathbf{q}$ (see Chapt. 3).

Consider the dynamics of our system inside a small vicinity of the instability threshold $(K > K_{th})$ where the spatial homogeneous solution becomes unstable and symmetry breaks down.

The excitation threshold is determined by [Fir]

$$K_{th}(q) = [2\sin(q^2 L/k)]^{-1}, \tag{11}$$

yielding the instability balloons shown in Fig. 6.

Early in the development of the system's dynamics when mode amplitudes are small enough the rolls' amplitudes grow exponentially, $a_q(t) \sim e^{\lambda_q t}$. The growth rate λ_q is proportional to $\mid K - K_{th} \mid$. This means the highest growth rates have rolls (modes) with wavevectors $\mathbf{q}_n^b$ located at the very bottom of the instability balloons, and become the so-called superactive (SA) modes (Fig. 6). Interaction between rolls is realized in the form of "Winner Takes All" (WTA) dynamics [VF]. Due to this intermode competition only superactive modes survive, while others die.

For modes with equal λ_q, the one with the advantage in initial conditions will become the SA mode winner (Fig. 7a). If growth rates are different, victory of the mode competition will go to the mode having the highest growth rate not dependent on initial conditions (Fig. 7b). This intermode competition mechanism results in the survival of only SA modes.

The next stage of system dynamics is hexagon formation [DAF, CCL]. A perfect hexagon is a system of three SA rolls with wavevectors $\mathbf{q}_1, \mathbf{q}_2$, and $\mathbf{q}_3$

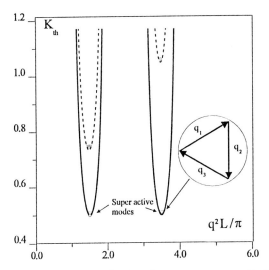

Fig. 6. Kerr-slice/feedback-mirror system. Roll excitation threshold curves versus diffraction parameter q^2L for a system without diffusion (solid lines), and with diffusion (dotted curves).

satisfying the resonance condition. Hexagons result from the interaction between superactive rolls having wavevectors $\mathbf{q}_1, \mathbf{q}_2, \mathbf{q}_3$ oriented at $\pi/3$ with respect to each other, as well as synchronized phases. The resonance condition for wavevectors is

$$\mathbf{q}_1 + \mathbf{q}_2 + \mathbf{q}_3 = 0. \tag{12}$$

It is easy to see that due to nonlinear interaction between two superactive rolls, only SA modes with wavevectors $\mathbf{q}_1$ and $\mathbf{q}_2$ oriented at $\pi/3$ can again give birth to SA mode $\mathbf{q}_3$ (Fig. 6). The three rolls forming a hexagon support each other; that is, inside the hexagon family we have so-called cooperative mode dynamics [Hir]. During inter-roll competition these three rolls either live or die together. They share common life, and must be considered as a unit dynamic object. Among hexagons with different wavevector orientations there is also inter-hexagon competition, and to be the winning hexagon the hexagon family must have some advantage in initial conditions [VF].

10 diffraction and Diffraction

Hexagonal patterns are rather common in the nonlinear dynamics of spatially extended systems: Rayleigh-Bénard convection in fluids [Sir], Bernard-Marangony convection [Kos], hexagonal Turing structures in chemical reactions [Kur], and finally, optical hexagons [GBV, TBW, KBT, Vor, BHG].

The optical branch of "hexagonal science" has some peculiarities related to the multibranch character of the instability area (a tribute to the specific local interactions caused by diffraction).

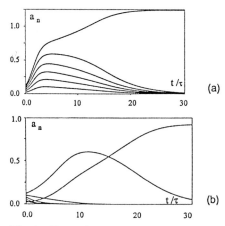

Fig. 7. Typical example of inter-mode competition in the Kerr-slice/feedback-mirror system [VF]. Evolution of six mode amplitudes in the pattern formation process. (a) SA modes with equal values for growth rate λ_q and different initial conditions. The mode winner has the largest value for the initial amplitude. (b) Modes with slightly different growth rates. The mode winner has the largest growth rate value independent of the initial conditions.

We must distinguish between *d*iffraction and *D*iffraction. *d*iffraction (with lowercase *d*) is related to the local scale of the problem, or diffraction on the self-induced phase inhomogeneities. *D*iffraction (with uppercase *D*) is related to the global effect, which is diffraction on the Kerr slice aperture[7].

*D*iffraction demonstrates the influence of boundary conditions on pattern formation. Boundary conditions are very important both in general and in particular for nonlinear problems. We will illustrate such problems using hexagonal patterns as an example.

First, note that expression (11) for instability branches does not consider the Kerr slice aperture size a. All pattern-formation characteristics depend only on the intrinsic properties of the system. This is a typical example of *d*iffraction.

All instability balloons shown in Fig. 6 have the same values for the "bottom-level balloon" $K_{th}(q_n^b) = 0.5$, where $q_n^b = [(k/L)(3\pi/2 + 2\pi n)]^{1/2}$; $(n = 0, 1, \cdots)$. Clearly, there is no spatial frequency cutoff and thus the smallest spatial scale of the problem doesn't exist. Remember, we were trying to ignore diffusion. This means the problem statement is not correct. What we have is degeneracy.

There are two ways to circumvent this degeneracy. Either: (1) account for the influence of diffusion (even if the diffusion length l_D is very small), or (2) consider weak diffractive effects on the Kerr-slice aperture (the "weak aperture effect"). In the first case, there is a natural spatial frequency cutoff related to the diffusion. Correcting (11) to account for diffusion we have [Fir]

$$K_{th}^D(q) = (1 + Dq^2) \cdot K_{th}(q). \tag{13}$$

---

[7] We develop here the idea begun by P. W. Anderson, who observed there are two types of broken symmetry, *b*roken and *B*roken [And].

Even weak diffusion does not create equal rights among instability balloons (Fig. 6, dotted balloons). For any $K > K_{th}$ we can find the balloon with the highest possible spatial frequency of mode excitation as determined by diffusion. Due to "inter-balloon" interactions, hexagonal patterns should be located at the bottom of the lowest instability balloon, which has the smallest spatial frequency q_0^b. (We consider the case of so-called "soft excitation," small initial phase modulation and small input field phase distortions.)

The influence of the weak aperture effect (weak *Diffraction*) is similar to diffusion. The high-order balloons go up, destroying the degeneracy. The influence of *Diffraction* first begins to appear for rolls with wavevectors $q > q_d = ka/L$, increasing their excitation threshold [VF]. This influence appears only in a narrow [approximately $(\lambda L)^{1/2}$] area near the Kerr slice aperture boundary. In the center of the aperture, the hexagonal pattern doesn't change very much and it is determined by the local properties of the system. But at the periphery of the aperture defects appear, destroying the perfect hexagonal structure. These defects mark the first appearance of the boundary effect, that is, the influence of *Diffraction*.

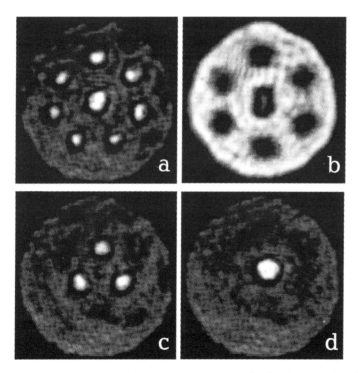

Fig. 8. Diffractive polygon-type patterns obtained using the 2-D feedback system shown in Fig. 2*b* (without block G), for different diaphragm diameters d located in the lens L_1 focal plane. (a) $d = 2.1mm$, (b) $d = 1.8mm$, (c) $d = 1.2mm$, (d) $d = 0.8mm$. The intensity reversal in (b) is related to uniform phase shift [ALV].

Radical changes occur when the radius of aperture diffraction influence $(\lambda L)^{1/2}$ is approximately equal to the aperture size so that $(\lambda L)^{1/2} \sim a$ or $L \sim L_d/\pi$, where $L_d = ka^2/2$ is diffractive length.

We can use wavevector language to make the same estimation. Diffraction on the aperture causes global changes if the aperture-determined wave number $q_d = ka/L$ has the same value as the characteristic spatial frequency of a perfect hexagonal pattern q_1^b, which gives a similar result: $L \sim (4/3)L_d/\pi$. For $L \ll L_d/\pi$, we deal with weak-aperture diffraction, or with diffraction. For $L = L_d/\pi$, Diffraction occurs [PAO]. Effect of the boundary conditions (Diffraction) is demonstrated in Fig. 8. Diffraction causes global changes of excited patterns, and instead of hexagons or the coexistence of hexagons and rolls we have different polygon-type patterns (see Fig. 8).

11 Far Away from Hexagons: Delay in Time and Space

Regardless of whether it is theory, calculation, or experiment, now nobody is surprised to obtain rolls/hexagons, or perhaps even various pleasing defects in hexagonal patterns [Gun, TR, ST, RGB, PN]. More likely, we are surprised by the lack of hexagon/roll appearance. The problem now is how to avoid or to kill hexagonal structures (it all depends on the nature of the researcher) and to discover something more sophisticated.

In fact, we have discussed some possibilities of "nonhexagonal" dynamics – multicomponent optical reaction-diffusion systems (Fig. 5). Examples of other types of patterns, including diffractive autosolitons, optical vortices and traveling-wave patterns, are presented in [RK] and Chapts. 5 and 7. Vortices and traveling-wave patterns add complexity to system dynamics through several dynamic variables having different time constants. In this case, we deal again with multicomponent system dynamics.

Another way to keep simple one-component mathematical models and at the same time obtain nontrivial dynamics is to add nonlocal interactions. The dynamic system with time delay is a good example of complex solution generation from trivial (no time-delay) equations (see [IDA, LeB] and Chapt. 4). Field rotation in a nonlinear optical system's 2-D feedback loop (i.e., OK-system) as suggested in [VIS, VKS], is, in some sense, an extension of time-delay ideology into spatial variables. This extension gives rise to a large variety of patterns, even in the case of diffusion-type transverse interactions (see Fig. 4). The analogy to time-delay systems is most conspicuous for the case of 1-D feedback systems with field rotation [ILV].

Formally, the transition to the one-dimensional case can be accomplished if we consider that in the function $u(\mathbf{r},t)$, describing nonlinear phase modulation the radial component of vector $\mathbf{r} = \{r, \theta\}$ is fixed: $r = r_0 = const.$[8] From the OK-equation (8) for the one-dimensional case with field rotation we obtain

---

[8] To provide 1-D feedback, masks in the form of thin transparent rings having different radii r_0 were used in the 2-D feedback OK-system shown in Fig. 2.

[VKS, AhI]

$$\tau \frac{\partial u(\theta, t)}{\partial t} + u(\theta, t) = D^\star \frac{\partial^2 u(\theta, t)}{\partial \theta^2} + R\{1 + \gamma \cos[u(\theta + \Delta, t) + \phi_0]\}, \qquad (14)$$

where $u(\theta, t) = u(r_0, \theta, t)$, Δ is angle of field rotation, and $D^\star = D/r_0^2$ is the effective diffusion coefficient.

The angle of rotation Δ plays the role of "spatial delay." The system's behavior is governed by the periodic boundary conditions

$$u(\theta, t) = u(\theta + 2\pi, t), \quad \frac{\partial u(\theta, t)}{\partial \theta} = \frac{\partial u(\theta + 2\pi, t)}{\partial \theta}.$$

The system's dynamics represent the elementary wave oscillators – rotating waves with different discrete wavevectors $q_n = n$, $(n = 1, 2, \cdots)$, and frequencies $\omega_n \tau = (1 + D^\star q_n^2) \tan(q_n \theta)$. Note that periodic boundary conditions are responsible for discrete spatial spectrum formation. To emphasize the importance of boundary conditions on dynamic behavior, we can identify the system with "spatial-delay" as a system with Diffusion.[9]

The solution of (14) can be represented as a sum of harmonic wave oscillators

$$u(\theta, t) = \sum_{n=1}^{N} a_n(t) \cos(\omega_n t + q_n \theta), \qquad (15)$$

where $a_n(t)$ are the mode amplitudes. Nonlinear analysis of (14) based on the Neumann's series approach shows that in the vicinity of a bifurcation point there is strong WTA-type competition among rotating modes, which provides a victory for only one mode, that mode having the largest eigenvalue (superactive modes) and the advantage in initial conditions [IV]. The presence of WTA-type mode dynamics, a style now popular in artificial neural network theory [SGr], gives a foundation (or rather a basis for speculation) for using continuously distributed nonlinear optical systems as a neural-network type of computer (neuromorphical optical systems (see [Ama, BA, Vor] and Chapts. 3 and 6).

Results of 1-D rotatory-wave theory as based on (14) and experimental data obtained from OK-type systems using an LCLV are in good qualitative agreement (see Figs. 9 and 10 in Chapt. 1 and [AhI, AkV, ILV]). Besides verification of

---

[9] Strictly speaking, all games with D and d have no meaning from a mathematical point of view as solutions of all partial differential equations depend both on the type of equation and on boundary conditions. There is no basis for arguing which is more important for the solution, equation or boundary conditions. From a physical point of view, there is sometimes perhaps a reason to separate an equation from the boundary conditions. In fact, this is one way of substituting one type of boundary condition for another for which there probably can be found a solution. But instead of confessing what they really have done, physicists prefer to hide these techniques with philosophical discussions about "internal" and "external" instability, breaking and retention of symmetry, and the like. Perhaps all of these magic words help physicists find a solution for situations mathematicians view as hopeless, such as when there are no theorems in existence or unique solutions.

the theory of elementary wave oscillators, this qualitative agreement lends support to the general validity of the Kerr-slice model for describing LCLV-based nonlinear optical systems.

One-dimensional wave oscillators in the system with "spatial delay" are an excellent example of complicated dynamics in a simple system. We mention several useful properties concerning this subject of nonlinear dynamics:

- There is only one simple-looking equation.
- Many modes can exist with one set of parameters [VR].
- Both hysteresis and strong intermode interactions occur – a good base for building nonlinear theory.
- We have experimental implementation of a wave oscillator model with a set of control parameters $(K, \Delta, D^\star, \gamma)$ that can be experimentally varied over a wide range.
- There are still possibilities for future development of this model, including interactions between wave oscillators with diffusion coupling (associative memory based on rotating waves), wave oscillators under the influence of external harmonic forces (resonance and parametric excitation of rotating waves), and 2-D rotating waves (see Fig. 3) for which the theory is still a challenge for researchers [AF, ZLV, Raz].

12 Diffusion + Diffraction + (Interference) + Nonlocal Interactions = Akhseals

This formula describes the physical phenomena that contribute to the formation of Akhseals, a type of pattern that looks like spectacular flowers shown in Figs. 9 and 10.[10]

---

[10] This is the story of the birth of the word "Akhseal." My memories are these: Moscow, last week of May 1991, blooming apple-tree gardens around the University, late night, and the last experiments with the OK-system are being done before summer vacation. These were not real experiments, but more like games – "Let us try this, or maybe this...." Fantastic patterns suddenly began to appear. These patterns have their own life – slow rotation of one pattern for a second, then a fast cascade of bifurcations. The old pattern dies, giving life over again only for a short time. Optical patterns replace each other, sometimes returning to patterns that we have already seen, somehow recreating the most remarkable patterns of all. The optical system is alive, and we are the only viewers of this fantastic movie. A small change in a parameter, and a different movie appears. Then, the door opens and Prof. S. A. Akhmanov enters the lab. We watch the "nonlinear optical" movie together in a dark, silent room. "All my life," he said, "I've been trying to show the beauty of nonlinear optics, but I never suspected it could be so spectacular." One month later, Prof. Akhmanov died after unsuccessful surgery. *Akhseal* = (*Akh*)manov (*Se*)rgey (*Al*)exandrovich is the name we have given to these patterns in memory of the person who shared with so many others the beauty of nonlinear optics. On September 1991 in St.Petersburg, during a special session dedicated to Prof. Akhmanov, we showed a movie about these optical patterns – Akhseals (see Proc. of this session [Vrt]). It was during this session the idea for this book first occurred. – M. A. Vorontsov

Fig. 9. Periodically alternating Akhseal-type patterns in the OK-system shown in Fig. 2*b* (block *G* is a field rotator) [vrt, TR].

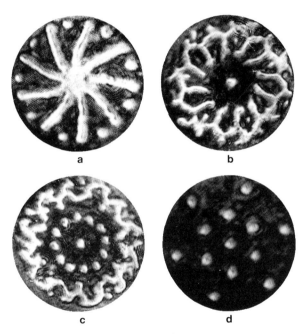

Fig. 10. Typical examples of Akhseal-type patterns in the OK-system. (a) - (d) correspond to different angles of field rotation Δ and feedback loop length L [Vrt].

How do Akhseals appear? Consider the OK-system schemes shown in Fig. 2. The pair of lenses (L_1, L_2) in Fig. 2 were used to avoid diffraction and to provide a poor diffusion approach. We do not need this anymore for Akhseal formation. Similar to hexagonal patterns Akhseals are also diffractive patterns, but different from those produced in systems with field rotation (block *G* in Fig. 2 is a field rotator). To increase diffraction, we need either to make the feedback loop longer, or to decrease the incident laser beam's diameter. Akhseals were observed both in the OK-system using the polarization effect (Fig. 2*b*), and in the OK-system with a single laser beam (reference mirror in Fig. 2a was removed).[11]

The OK-system mathematical model for diffraction and field rotation involves two equations [Vr]. The first is the OK-type equation (8) for nonlinear phase modulation,

$$\tau\frac{\partial u(r,\theta,t)}{\partial t} + u(r,\theta,t) = D\nabla_\perp^2 u(r,\theta,t) + K\mid A_{FB}(r,\theta+\varDelta,t)\mid. \qquad (16)$$

This equation describes wave propagation in the feedback loop (diffraction over a distance L),

$$-2ik\frac{\partial A}{\partial z} = \nabla_\perp^2 A, \qquad (17)$$

with the boundary conditions (2) determined in the LCLV plane

$$A(\mathbf{r},0,t) = I_0^{1/2}\exp\left[iu(\mathbf{r},t)\right], \quad (\mathbf{r}=\{r,\theta\}). \qquad (18)$$

The expression for feedback intensity on the right-hand side of (16) is different for the OK-system with and without interference

$$A_{FB} = A(\mathbf{r},L,t) + \gamma I_0^{1/2}\exp(iu_0). \qquad (19)$$

For the OK-system with poor diffractive effect $\gamma = 0$.

Conditions for Akhseal excitation, and direct comparison of theoretical results with numerical simulations and experimental results, were done only for specific rotation angles of the feedback field $\varDelta = 2\pi/N$ [Vr, Vnt]. In this case, an Akhseal is formed from N SA modes (rolls) with wavevectors $\mathbf{q}_n$ that satisfy the resonance conditions similar to (12),

$$\mathbf{q}_1 + \mathbf{q}_2 + \cdots + \mathbf{q}_N = 0, \qquad (20)$$

$$\mid \mathbf{q}_n \mid = \mid \mathbf{q}^b \mid = \kappa,$$

where κ is the wave number of the most unstable modes, with wavevectors located at the very bottom of the instability balloon. Amplitudes of modes $a_n(t)$ forming the Akhseal family satisfy the following system of equations obtained from linear stability analysis [Vr]:

$$\tau\frac{da_1}{dt} + (1+\kappa^2 D)a_1 = 2K\sin(\kappa^2 L)a_2,$$

$$\tau\frac{da_2}{dt} + (1+\kappa^2 D)a_2 = 2K\sin(\kappa^2 L)a_3,$$

$$\vdots$$

$$\tau\frac{da_{N-1}}{dt} + (1+\kappa^2 D)a_{N-1} = 2K\sin(\kappa^2 L)a_N,$$

$$\tau\frac{da_N}{dt} + (1+\kappa^2 D)a_N = 2K\sin(\kappa^2 L)a_1, \qquad (21)$$

---

[11] In the title of this section we use the word *interference* in brackets to emphasize that Akhseals can be observed in systems with poor diffractive effect.

The system of mode equations (21) may be assigned to a differential equation with cooperative dynamics – all N modes live or die together [BA, Hir]. As well as WTA-dynamics, this type of dynamic mode behavior in nonlinear optical systems shows a high degree of similarity to the basic mechanisms of artificial neural networks [SGr, AMP].

Compare the processes of hexagon and Akhseal-type pattern formation. We can point to at least two significant distinctions. First, by altering the angle $\Delta = 2\pi/N$, we can control the number of modes participating in pattern formation. In the case of hexagonal patterns, the number of modes is always three. Second, the interaction between SA modes required to form hexagons manifests itself only through an increase in second-order terms in the feedback intensity expansion, that is, only if the mode amplitudes are big enough. Akhseal-type patterns appear due to linear (first-order) terms. For this reason, Akhseals are more robust structures and are more independent of the influence of incident field phase distortions and external noise than are hexagonal patterns.

Akhseal dynamics behave differently for even and odd N. For even N, the system of instability balloons for Akhseals is coincident with the corresponding instability balloons of hexagonal patterns in the Kerr-slice/feedback-mirror system shown in Fig. 6. The one difference is that with Akhseals there are no gaps between instability balloons [Vnt].

The most complicated and interesting dynamics occur for odd N. When diffusion coefficient D is small, the most unstable balloon becomes the second instability balloon, so that $\kappa = q_2^b$.

The second distinguishing feature of Akhseal patterns is that for odd N mode amplitudes $a_n(t), n = 1, \cdots, N$ forming Akhseals may oscillate with the same frequency ω for all N modes as determined by expression [Vnt]:

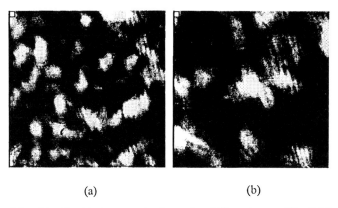

(a)   (b)

Fig. 11. Spatio-temporal chaos in OK-system with field rotation. Typical output-intensity patterns. Spatial filtering in the feedback loop allows us to control the spatial statistical properties of resulting structures. Both (a) and b) correspond to different spatial filters [VRC].

$$\tau\omega = (1 + \kappa^2 D)\tan(\Delta/2).$$

The presence of mode amplitude oscillations corresponds to the appearance of rotating structures.

The theory of Akhseal-type patterns is now at its very beginning. The patterns we have shown in this introduction give us only a limited idea of the fantastically rich dynamics of Akhseals. The great variety of pattern formation regimes shown (and not shown) in Figs. 9 and 10 occur just within a small vicinity of the excitation threshold.

When incident light intensity is increased pattern disintegration begins, leading us into a new and different world – spatio-temporal chaos (Fig. 11).

References

[NP]   Nicolis, G., and Prigogine, I.: Self-Organization in Non-Equilibrium Systems. Wiley, New York, 1977.

[MY]   Monin, A.S., and Yaglom, A.M.: Statistical Fluid Mechanics. MIT Press, Cambridge, MA, London, 1975.

[Hak]   Haken, H.: Synergetics. Springer-Verlag, Berlin, Heidelberg, 1978.

[CH]   Cross, M.C., and Hohenberg, P.C.: Pattern Formation Outside of Equilibrium. Rev. Mod. Phys. 65 **3** 1993, 851.

[Sil]   Bar-Joseph, I., and Silberberg, Y.: Instability of Counterpropagating Beams in a Two-Level-Atom Medium. Phys. Rev. A **36** 1987, 1731.

[LRT]   Le Berre, M., Ressayre, E., and Tallet, A.: Self-Oscillations of the Mirrorlike Sodium Vapor Driven by Counterpropagating Light Beams. Phys. Rev. A **43** 1991, 6345.

[SM]   Ségard, B., and Macke, B.: Self-Pulsing in Intrinsic Optical Bistability with Two-Level Molecules. PRL **60** 1988, 412.

[GVK]   Giusfredi, G., Valley, J.F., Khitrova, G., and Gibbs, H.M.: Optical Instabilities in Sodium Vapor. J. Opt. Soc. Am. A, 5 **5** May 1988, 1181.

[LOT]   Lugiato, L.A., Oppo, G.L., Tredicce, J.R., Narducci, L.M., and Pernigo, M.A.: Instabilities and Spatial Complexity in a Laser. J. Opt. Soc. Am. B, **7** 1990, 1019.

[Lug]   Lugiato, L.A.: Spatio-Temporal Structures, Part 1. Phys. Rept. 3-6 **219** 1992.

[Wei]   Weiss, C.O.: Spatio-Temporal Structures, Part 2, Vortices and Defects in Lasers. Phys. Rept. 3-6 **219** 1992, 311.

[MN]   Moloney, J.V., and Newell, A.C.: Nonlinear Optics. Addison-Wesley, Redwood City, CA, 1991.

[FH]   Fuchs, A., and Haken, H.: Neural and Synergetic Computers. Springer-Verlag, Berlin, 1988.

[Shn]   Shen, I.R.: The Principles of Nonlinear Optics. J. Wiley & Sons, New-York, 1984.

[OI]   Otsuka, K., and Ikeda, K.: Cooperative Dynamics and Functions in a Collective Nonlinear Optical Element System. Phys. Rev. A 10 **39** 1989, 5209.

[Fir]   Firth, W.J.: Spatial Instabilities in a Kerr Medium with Single Feedback Mirror. J. Mod. Opt. **37** 1990, 151.

[DF]   D'Alessandro, G. and Firth, W.J.: Spontaneous Hexagon Formation in a Nonlinear Optical Medium with Feedback Mirror. Phys. Rev. Lett. **66** 1991, 2597.

[VF]   Vorontsov, M.A., and Firth, W.J.: Pattern Formation and Competition in Non-
        linear Optical Systems with Two-Dimensional Feedback. Phys. Rev. A 4 **49**
        April 1994, 2891.

[CG]   Courtois, J.Y., and Grynberg, B.: Spatial Pattern Formation for Counter-
        Propagating Beams in a Kerr Medium: A Simple Model. Opt. Comm. **87** 1992,
        186.

[GB]   Gaeta, A.L., and Boyd, R.W.: Transverse Instabilities in the Polarizations and
        Intensities of Counterpropagating Light Waves. Phys. Rev. A 2 **48** 1993, 1610.

[Mur]   Murray, G.: Lectures on Nonlinear Differential Equations Models in Biology.
        Clarendon Press, Oxford, 1977.

[KPP]   Kolmogorov, A.N., Petrovskii, I.G., and Piskunov, N.S.: Vestn. Mosk. Univ.
        Mat. Mekh 1 **1** 1937.

[FG]   Firth, W.J., and Galbraith, I.: Diffusive Transverse Coupling of Bistable El-
        ements – Switching Waves and Crosstalk. IEEE J. Quant. Elect. 9 **21** 1985,
        1399.

[ML]   Michailov, A.S., and Loscutov, A. Yu.: Foundations of Synergetics. Springer-
        Verlag, Heidelberg, Berlin, 1991.

[MZI]   Vorontsov, M.A., Zheleznykh, N.I., and Ivanov, V.Yu.: Transverse Interactions
        in 2-D Feedback Non-Linear Optical Systems. Opt. & Quant. Elect. **22** 1990,
        501.

[RRH]   Rzhanov, Yu.A., Richardson, H., Hagberg, A.A., and Moloney, J.V.: Spa-
        tiotemporal Oscillations in a Semiconductor Etalon. Phys. Rev. A 2 **4** 1993,
        1480.

[Ros]   Rosanov, N.N.: Spatial Effects in Bistable Optical Systems. In New Physi-
        cal Principles of Information Optical Processing. Eds. Akhmanov, S.A., and
        Vorontsov, M.A., Moscow, Nauka, 1989, 230.

[VDP]   Vorontsov, M.A., Dumarevsky, Yu.D, Pruidze, D.V., and Shmalhauzen, V.I.:
        Auto-Wave Processes in the Systems with Optical Feedback. Izv. AN USSR
        Fiz. 2 **52** 1988, 374.

[VPS]   Vorontsov, M.S., Pruidze, D.V., and Shmalhauzen, V.I.: Spatial Bistability in
        Nonlinear Optical System with Optical Feedback. Izvestiya Vyssh. Uchebn.
        Zaved. Radiofizika 12 **25** 1988, 505.

[Vor]   Vorontsov, M.A.: Problems of Large Neurodynamics System Modelling: Op-
        tical Synergetics and Neural Networks. Proc. Soc. Photo-Opt. Instrum. Eng.
        1402 1991, 116.

[AVL]   Akhmanov, S.A., Vorontsov, M.A., Ivanov, V.Yu., Larichev, A.V., and
        Zheleznykh, N. I.: Controlling Transverse-Wave Interactions in Nonlinear Op-
        tics: Generation and Interaction of Spatiotemporal Structures. J. Opt. Soc.
        Am. B 1 **9** January 1992, 78.

[VKS]   Vorontsov, M.A., Koriabin, A.V., and Shmalhauzen, V.I.: Controlling Optical
        Systems. Nauka, Moscow, 1988.

[VoL]   Vorontsov, M.A., and Larichev, A.V.: Adaptive Compensation of Phase Distor-
        tions in Nonlinear System with 2-D Feedback. Proc. Soc. Photo-Opt. Instrum.
        Eng. **1409** 1991, 260.

[ALV]   Arecchi, F.T., Larichev, A.V., and Vorontsov, M.A.: Polygon Pattern Forma-
        tion in a Nonlinear Optical System with 2-D Feedback. Opt. Comm. **105** 1994,
        297.

[VIS]   Vorontsov, M.A., Ivanov, V.Yu., and Shmalhausen, V.I.: In Laser Optics of
        Condensed Matter: Proc. 3rd Binational USA-USSR Symp. Plenum Publish-
        ing, New York, NY, 1988, 507.

[AhI]   Akhmanov, S.A., Vorontsov, M.A., and Ivanov. V.Yu.: Large-Scale Transverse
         Nonlinear Interactions in Laser Beams; New Types of Nonlinear Waves. JETP
         Lett. 12 **47** 1988, 707.

[AkV]   Akhmanov, S.A., Vorontsov, M.A., and Larichev, A.V.: Dynamics of Nonlinear
         Rotating Light Waves: Hysteresis and Interaction of Wave Structures. Sov. J.
         Quant. Elec. 4 **20** Apr. 1990, 325.

[VIL]   Vorontsov, M.A., Ivanov, V.Yu., and Larichev, A.V.: Rotatory Transversal
         Instability of Laser Field in Nonlinear Systems with Optical Feedback. Izv.
         AN USSR Fiz. 2 **55** 1991, 316.

[ILV]   Ivanov, V.Yu., Larichev, A.V., and Vorontsov, M.A.: 1-D Rotatory Waves in
         the Optical System with Nonlinear Large-Scale Field Interactions. Proc. Soc.
         Photo-Opt. Instr. Eng. **1402** 1991, 145.

[TNT]   Thuring, B., Neubecker, R., and Tschudi, T.: Opt. Comm. **102** 1993, 111.

[PRR]   Pampaloni, E., Ramazza, P.-L., Residori, S., and Arecchi, F. T.: Experimental
         Evidence of Boundary Induced Symmetries in an Optical System with Kerr-
         Like Nonlinearity. Europhys. Lett. **25** 1994, 587.

[PRA]   Pampaloni, E., Residori, S. and Arecchi, F.T.: Roll-Hexagon Transition in a
         Kerr-Like Experiment. Europhys. Lett. **24** 1993, 647.

[HLR]   Hide, R., Lewis, S.R., and Read, P.L.: Sloping Convection: A Paradigm for
         Large-Scale Waves and Eddies in Planetary Atmospheres Phys. Rev. 2 **4** 1994,
         135.

[DDK]   Danilov, S.D., Dolzhanskii, F.V., and Krymov, V.A.: Quasi-Two-Dimensional
         Hydrodynamics and Problems of Two-Dimensional Turbulence. Phys. Rev. 2
         4 1994, 299.

[Are]   Arecchi, F.T.: Space-Time Complexity in Nonlinear Optics. Physica D **51** 1991,
         450.

[Sta]   Staliunas, K.: Laser Ginzburg-Landau Equation and Laser Hydrodynamics.
         Phys. Rev. A 2 **48** 1993, 1573.

[Hkn]   Haken, H.: Advanced Synergetics. Springer-Verlag, New York, Berlin, 1987.

[Kur]   Kuramoto, Y.: Chemical Oscillations, Waves and Turbulence. Springer-Verlag,
         Berlin, Heidelberg, New York, Tokyo, 1984.

[DV]   Degtyarev, E.V., and Vorontsov, M.A.: Spatial Dynamics of a Two-Component
         Optical System with Large-Scale Interactions. Mol. Cryst. Liq. Cryst. Sci.
         Technol., Sec. B: Nonlinear Optics, **3** 1992, 295.

[VRC]   Vasiliev, V.A., Romanovski, Yu.M., Chernavsky, D.S., and Jakhno, V.G.:
         Autowave Processes in Kinetic Systems. VEB Deutscher Verlag der Wis-
         senschaften, Berlin, 1987.

[RVP]   Rakhmanov, A.N, Vorontsov, M.A., Popova, A.V., and Shmal'gauzen, V.I.:
         Optical Model of a Two Dimensional Active Medium. Sov. J. Quant. Elect. 7
         22 July 1991, 593.

[VRS]   Vorontsov, M.A., Rakhmanov, A.N., and Shmal'gauzen, V.I.: Optical Self-
         Oscillatory Medium Based on a Fabry-Pérot Interferometer. Sov. J. Quant.
         Elect. 1 **22** Jan. 1992, 56.

[Vrt]   Vorontsov, M.A.: Nonlinear Spatial Dynamics of Light Field. Izv. AN USSR
         Fiz. 4 **56** 1992, 7.

[DAF]   D'Alessandro, G., and Firth, W.J.: Hexagonal Spatial Patterns for a Kerr Slice
         with a Feedback Mirror. Phys. Rev. A 1 **46** 1992, 537.

[CCL]   Ciliberto, S., Coullet, P., Lega, J., Pampaloni, E., and Perez-Garcia, C.: Defects
         in Roll-Hexagon Competition. Phys. Rev. Lett. 19 **65** 1990, 230.

[Hir]   Hirsch, M.W.: Systems of Differential Equations that Are Competitive or Co-
        operative. SIAM, J. Math. Analy. **16** 1985, 423.
[Sir]   Sirovich, L.: Chaotic Dynamics of Coherent Structures. Physica D **37** 1989,
        126.
[Kos]   Koschmieder, E.L.: Bénard-Cells and Taylor Vortices. Cambridge U.P., Lon-
        don, 1992.
[GBV]   Grynberg, H. Le Bihan, E., Verkerk, P., Simoneau, P., Leite, J.R.R., Bloch, D.,
        Le Boiteux, S., and Ducloy, M.: Observation of Instabilities Due to Mirrorless
        Four-Wave Mixing Oscillation in Sodium. Opt. Comm. **67** 1988, 363.
[TBW]   Tamburrini, M., Bonavita, M., Wabnitz, S., and Santamato, E.: Hexagonally
        Patterned Beam Filamentation in a Thin Liquid Crystal Film with a Single
        Feedback Mirror. Opt. Lett. 11 **18**, 1993.
[KBT]   Kreuzer, M., Balzer, W., and Tschudi, T.: Formation of Spatial Structures in
        Bistable Optical Elements Containing Nematic Liquid Crystals. Opt. Lett. 4
        29 February 1990, 579.
[BHG]   Banerjee, P.P., Hseuh-Lin Yu, Gregory, D., Kukhtarev, N., and Caulfield, H.J.:
        Self-Organization of Scattering in Photorefractive KNbO (3) into a Reconfig-
        urable Hexagonal Spot Array. Opt. Lett., 1994.
[And]   Anderson, W.P.: Science, **177** 1972, 393.
[PAO]   Papoff, F., D'Alessandro, G., Oppo, G.-L., and Firth, W.J.: Local and Global
        Effects of Boundaries on Optical Pattern Formation in Kerr Media. Phys. Rev.
        A, June 1993.
[Gun]   Gunaratne, G.: Complex Spatial Patterns on Planar Contiinua. Phys. Rev.
        Lett. 9 **71** Aug. 1993, 1367.
[TR]   Tsimring, L.S., and Rabinovich, M.I., Dislocations in Hexagonal Patterns. In
        Spatio-Temporal Patterns in Nonequlibrium Complex Systems. Report, NATO
        Advanced Research Workshop, Santa Fe, April 1993.
[ST]   Sushchik, M.M., and Tsimring, L.S.: The Eckhaus Instability in Hexagonal
        Patterns. Physica D **74** 1994, 90.
[RGB]   Rodriguez, J.M., Garcia, C.P., Bestehorn, M., Neufeld, M., and Frieddrich, R.:
        Motion of Defects in Rotating Fluids. Phys. Rev. 2 **4** 1994, 369.
[RK]   Rosanov, N.N. and Khodova, G.V.: Diffractive Autosoliton in Nonlinear Inter-
        ferometers. J. Opt. Soc. Am. B 7 **7** 1990, 1057.
[IDA]   Ikeda, K., Daido, H., and Akimoto, O.: Optical Turbulence: Chaotic Behavior
        of Transmitted Light from a Ring Cavity. Phys. Rev. Lett. **45** 1980, 709.
[LeB]   Le Berre, M., Ressayre, E., and Tallet, A.: Gain and Reflectivity Characteris-
        tics of Self-Oscillations in Self-Feedback and Delayed Feedback Devices. Opt.
        Comm. **87** 1992, 358.
[IV]   Iroshnikov, N.G., and Vorontsov, M.A.: Transverse Rotating Waves in the Non-
        linear Optical System with Spatial and Temporal Delay. In Essay in Nonlinear
        Optics: In Memoriam of Serge Akhmanov. Eds. Walther, H., and Koroteev,
        N.: M. Scully.-IOP, London, 1992.
[SGr]   Grossberg, S.: Nonlinear Neural Networks: Principles, Mechanisms and Archi-
        tectures. Neural Networks **1** 1988, 17.
[Ama]   Amari, S.: Field Theory of Self-Organizing Neural Nets. IEEE Trans. SMC **13**
        1983, 741.
[BA]   Benkert, C., and Anderson, D.Z.: Controlled Competitive Dynamics in a Pho-
        torefractive Ring Oscillator: "Winner-Takes-All" and "Voting-Paradox" Dy-
        namics. Phys. Rev. A, 7 **44** 1991, 4633.

[VR]   Vorontsov, M.A., and Razgulin, A.V.: Properties of a Global Attractor in a
          Nonlinear Optical System Having Nonlocal Interactions. Photonics and Opto-
          elect., 2 1 1993, 103.
[AF]   Adachihara, H., and Faid, H.: Two Dimensional Nonlinear – Interferometer
          Pattern Analysis and Decay of Spirals. J. Opt. Soc. Am. B, 7 10 1993, 1242.
[ZLV]   Zheleznykh, N.I., Larichev, A.V., and Vorontsov, M.A.: 2-D Dynamics of Neu-
          romorphic Optical System with Large-Scale Interactions. Proc. Soc. Photo-
          Opt. Instrum. Eng., 1402 1991, 154.
[Raz]   Razgulin, A.V.: Bifurcational Light Structures in Nonlinear Optical System
          with Nonlocal Interactions. Proc. Soc. Photo-Opt. Instr. Eng., 1961 1993,
          241.
[Vr]   Vorontsov, M.A.: "Akhseals" as a New Class of Spatio-Temporal Instabilities
          of Optical Fields. Sov. J. Quant. Elect. 23 1993, 269.
[Vnt]   Vorontsov, M.A., Iroshnikov, N.G., and Abernathy, R.L.: Diffractive Patterns
          in a Nonlinear Optical System with Field Rotation. In Chaos, Solitons and
          Fractals, 4. Ed. Lugiato, L, Pergamon Press, New York, 1994.
[AMP]   Abu-Mostafa, Y.S., and Psaltis, D.: Optical Neuron Computers. Sci. Amer. 3
          256 1987, 88.
[VRC]   Vorontsov, M.A., Ricklin, J.C., Carhart, G.W., Gose, D., and Miller, W.B.:
          Turbulent Phase Screen for Study of Imaging System Performance. J. Mod.
          Opt. to be published in 1994.
[PN]   Passot, T., and Newell, A.C.: Towards a Universal Theory of Natural Patterns.
          Physica D 74, 1994, 301.

1 Information Processing and Nonlinear Physics: From Video Pulses to Waves and Structures

S.A. Akhmanov

Lomonosov Moscow State University, Russia

Introduction

Prospects for an all-optical computer are governed entirely by the possibilities of nonlinear optics.

Advances in the development of efficient nonlinear materials and optical triggers have paved the way for development of ultrafast digital optical logic. Promising perspectives seem to have been opened up for the nonlinear optics modeling of neural networks, optical associative memory, and nonlinear analog processors.

Much should be expected from combining the methods of nonlinear optics and molecular electronics.[1]

1 Information Encoding by Carrier Modulation and the Physics of Nonlinear Oscillations and Waves

The principles of signal encoding, information processing, and operation of digital and analog processors where message signals interact are deeply rooted in nonlinear physics. Beginning in the early 1960s, the physics of nonlinear oscillations and nonlinear waves has been gaining in significance (see, e.g., [Wig, Got, Ste, AkR, Man, Pap, AK1, AK2, Lan]). Undoubtedly the core idea underlying this field of development was the one stated by von Neumann in the late 50s of information coding by means of the amplitude, phase, or frequency modulation of a carrier frequency (see [Wig] for a detailed account). Figure 1 illustrates representations of binary information (which are naturally extended to the case of multivalued logic), along with the traditional information by video pulses.

What were the advantages offered by the new way of data encoding? The first time, the main emphasis was placed on speedup of the trigger switching. Indeed, in those years, triggers operating with video pulses could be switched as fast as $10^{-7} - 10^{-8}$ s. Naturally, the transition to gigahertz carrier frequencies

---

[1] Ed. Note: This is the last article written by Prof. S. A. Akhmanov. It was published in *New Principles of Optical Computing*, eds. S. A. Akhmanov and M. A. Vorontsov, (in Russian) Nauka, Moscow, 1992.

($\omega_c \simeq 10^9 - 10^{10}$ Hz) called for the development of triggers with at least 100 times faster switching rates. [VPS].

An experimental realization of these ideas was reported before long. Goto [Got] was the first to point out that a parametrically excited resonant circuit made as a degenerate single loop parametric oscillator or parametron (classical object in the physics of nonlinear oscillations (see, e.g., [Man]) could be used as a binary phase trigger. Oscillations of the same amplitude and frequency phase shifted by π can be used to represent logical "0" and "1" (Fig. 1b). They are excited in a parametron, depending on the initial conditions or on the parameters of external forcing.

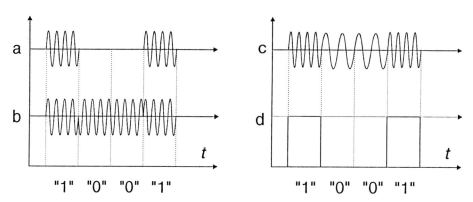

Fig. 1. Binary representation in systems with carrier frequency: (a) amplitude modulation, (b) phase modulation, (c) frequency modulation, (d) video pulses.

Efficient, small-size, radio frequency parametrons ($\omega_c \simeq 3-4$ MHz) were developed in several Japanese laboratories in the late 50s and early 60s. Computers were demonstrated that used phase data recording in both processor and main memory. At the megahertz carrier frequency, speed was not the main feature of this equipment. However, the RF parametrons exhibited high reliability and long life. New machine logic for such computers and new methods of data flow control were of great significance, too [Got, AkR]. It should be emphasized that by the early 60s the high switching speed of parametrons had been demonstrated and microwave parametrons with $\omega_c \simeq 2-4$ GHz had been developed (see [Ste]).

In those years, parametrons with three and more stable states were being developed and the principles of a multivalued parametron logic were formulated [AkR]. During the 60s, there was some interest in the amplitude and frequency coding of radio frequencies, but the work did not advance beyond the definition phase and simple demonstration experiments. The most advanced developments in this field, based on RF and microwave parametrons, lost the competition to semiconductor technology. Nevertheless, the upsurge of interest in the parametric devices associated with their potential application in digital computers has had far-reaching consequences. In discussing the physics and technology of data processing, we should point out two circumstances.

The principles of majority logic and the methods of organization of data flow in the system of RF self-excited oscillators – phase triggers, which had been developed for parametron-based digital computers, were extended to different hardware, specifically to computers with parametric Josephson components.

Extension of the carrier-modulation coding principle over systems in the microwave range involved the idea of a wave trigger. This is a nonlinear-wave system with discrete values for the phase or amplitude of a *traveling wave.*

Akhmanov and Khokhlov [AK2] have elaborated the theory of wave phase trigger. In this device, the phase of the traveling wave exhibits a pronounced bistable behavior. The phenomenon underlying this kind of phase bistability is the *phase selectivity* of a parametric traveling-wave amplifier. In this sense, the above wave trigger is the spatial analogue of the Goto parametron, though much faster – a typical switching time can be $0.1\omega_c^{-1}$. It was the first device of this type. Today the physics of nonlinear waves can offer a hierarchy of such systems.

It is worth noting that the peak of interest in the new applications of parametric systems was observed in the early years of optical quantum electronics. This explains why the early 60s saw the first propositions of light parametric amplifiers and oscillators. At the first stage of these studies, the main emphasis was laid on the purely optical problem of a tunable coherent light generator (a review of relevant papers may be found in [AK2]). In recent years considerable attention has been noted in the information aspect of parametric optics. The parametric interactions of light waves can be used to generate classical and quantum squeezed states of the light field.

Thus, the interaction and mutual penetration of the methods of information science and the physics of nonlinear interactions and waves have opened up new directions of research.

2 Modulation of Light Waves and Information Encoding in Digital Optical Computers. Optical Triggers

Digital and analog optical computers depend for their operation upon the modulation of a carrier frequency in much the same way as their RF counterparts do.[2] However, in this field the possibilities of optics far outperform the possibilities of radio physics. One can encode a light carrier by modulating its amplitude (intensity), frequency, phase, and polarization state. This can be done in the time domain with the characteristic time 10^{-13} to 10^{-14} s approaching the period of light oscillations, and in the spatial domain with the lowest scale of spatial oscillation approaching the wavelength.

The problem central to the physics and technology of an all-optical computer is the one of an efficient, fast optical switching device – an optical trigger. The

---

[2] In the literature, the name "optical computer" is often used to denote linear analog optical processors and various optoelectronic (hybrid) systems. In this text, by optical computer, we mean a computing system with a nonlinear digital or analog optical processor. This connotation is emphasized by the term "all-optical computer."

earliest propositions were reported soon after the advance of the laser in the early 60s. The complex nonlinear response of a multimode laser can bring about frequency bistability or multistability. A common property of gas lasers with a nonlinear absorber and semiconductor lasers is the amplitude bistability and a hysteresis cycle of the output characteristic curve. Clearly, these phenomena could be used to realize operations of digital logic. However, they have had no serious consequences for computer hardware. Laser optical triggers are rather sophisticated and consume too much power to be considered as serious candidates for optical computer components.

A new upsurge of interest in digital optical computers relates to the late 60s and early 70s. The fast progress of nonlinear optics made it possible to observe and use the amplitude optical bistability in a passive cavity, such as a Fabry-Pérot resonator, filled with a medium with cubic nonlinearity, i.e., with a medium exhibiting nonlinear absorption or refraction (see, e.g., [Szo, Lug]). The physics of this phenomenon is rather simple and easily tractable.

By way of example, consider a Fabry-Pérot resonator (Fig. 2) filled with a medium whose refractive index n depends on the incident intensity I by the law

$$n = n_0 + n_2 I. \tag{1}$$

The transparency $T = I_{out}/I_{in}$ of the resonator depends on the phase shift between the mirrors. In a cavity filled with a nonlinear medium, the total phase shift is a function of intensity

$$\Phi = \Phi_0 + \alpha I_{in} T, \tag{2}$$

where $\alpha \sim n_2$ is a constant. The output intensity can be found by substituting (2) in the familiar expression for cavity transparency

$$T = \frac{I_{out}}{I_{in}} = \frac{T_0}{1 + F \sin^2(\Phi/2)}. \tag{3}$$

Figure 2b illustrates a simple procedure for graphical solution of the nonlinear cavity equation. Drawing a straight line (2) with the slope $1/\alpha I_{in}$, locates an operating point on the transmission curve (3). Tracing the movement of the operation point for different input intensities yields the plot I_{out} versus I_{in} shown in Fig. 2c.

Amplitude optical bistability in a passive cavity filled with a medium of nonlinear refractive (dispersive optical bistability was first observed by McCall and coworkers ([McC]) in Na vapor (Fig. 3; cf Fig. 2c).

This was undoubtedly an important step toward a trigger amenable to optical logic: passive nonlinear resonators can be tiny devices with very low power requirement. This amplitude optical trigger is very similar to the RF amplitude trigger built around a resonance circuit with nonlinear capacitance or inductance.

The question here is essentially on the transfer in optics of the ideas of "time-domain" optical bistability elaborated first in the radio physics of lumped parameter systems. However, nonlinear optics brings about another, principally

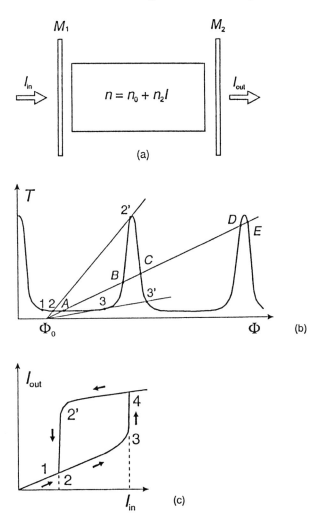

Fig. 2. Amplitude optical trigger using a nonlinear Fabry-Pérot cavity: (a) schematic representation, (b) graphical solution for the nonlinear cavity, (c) transmitted intensity is plotted against the incident intensity using the graphical solution method from (b).

different type of "spatial" bistability. Here the key problem is the method of feedback arrangement.

In the Fabry-Pérot cavity sketched in Fig. 2, the feedback loop is one-dimensional and dominated by nonlinear longitudinal interactions. The interactions of the field at different points over a cross section of the beam may be neglected.

The type of feedback can be essentially changed by incorporating the *transverse interactions* with controlled spatial scales and topology. This goal is achieved by introducing various spatial transformations (translation, tilting, rotation, extension, and compression) of the field in the feedback loops of the nonlinear ring resonator. A nonlinear resonator equipped with such a *two-dimensional*

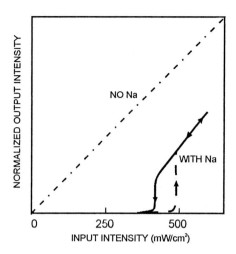

Fig. 3. Dispersive optical bistability in Na vapor, showing stable upper and lower states that could serve as "1" and "0" of an optical memory.

feedback (see [AkV]) reveals a number of new nonlinear wave phenomena, which have no remote analogues in nonlinear radio physics.

Figure 4 shows a ring cavity configuration with two-dimensional feedback that was designed to observe spatial optical bistability.

The field at beam cross-section points **r** and **r′** finds itself interrelated. If this cavity is filled with a medium with nonlinear refractive index (1), the spatial light distribution will exhibit bistable behavior and hysteresis.[3] The bifurcation diagram in Fig. 4b shows the quantities u_1 and u_2 proportional to the intensities at the labeled points of the beam cross section plotted versus the nonlinearity parameter $K = 2\pi n_2 I_{in} L/\lambda$, where L is the length of the nonlinear medium. The resultant output field reveals one of the two complementary stable configurations (alternating light and dark fields) representing spatial optical bistability. Just as with temporal bistability, the system can be switched from one stable state to the other by varying the initial conditions or by applying an external force.

Amplitude optical triggers with one- and two-dimensional feedback represented schematically in Figs. 2 and 4 can be used as basic elements of digital optical computers with different architecture. Modern nonlinear optics also offers a good possibility for the development of phase and polarization triggers.

Of course, it is a long way from a proposition of optical logic to a real optical computer. The first substantial assessment of the competitive strength of an all-optical computer using passive nonlinear optical elements was given by Keyes and Armstrong [KeA]. They drew a pessimistic conclusion: None of the existent nonlinear materials and available optical logic elements could afford a superior position with respect to chip technology in regard to the collection of

---

[3] Ed. Note: See also [VZI, AVL].

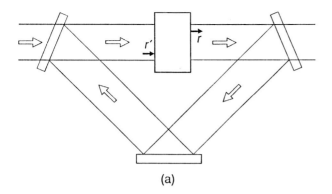

(a)

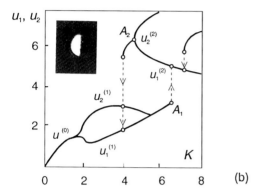

(b)

Fig. 4. Spatial optical bistability in a nonlinear resonator with two-dimensional feedback: (a) schematic of 2-D feedback arrangement, (b) discrete stable states correspond to variations of the spatial intensity distribution.

such parameters as speed, suitability for integration, power requirement, and heat evolution.

What is the state of the art to date? From the practical standpoint, it should be recognized that the most impressive progress has been made in the technology of amplitude optical triggers based on semiconductor microresonators. Figure 5 shows an array of bistable optical microresonators based on GaAs – AlAs superlattice [McJ].

The microresonators are configured as in the scheme depicted in Fig. 2. The logic OR arranged with such devices is switched by a light pulse with the energy $\omega_s \approx 2 \cdot 10^{-11}$ J. The characteristic switching time is about $\tau_s \approx 10^{-10}$ s. Optimization is expected to cut down the switching energy and time parameters to about $\omega_s \approx 2 \cdot 10^{-15}$ J (30 times the value of the fluctuation limit) and $\tau_s \approx 10^{-12}$ s, respectively. The authors claim a further improvement of these figures with the use of one-dimensional systems (such as polydiacetylene) or specifically designed organic molecules with very strong optical transitions as nonlinear media.

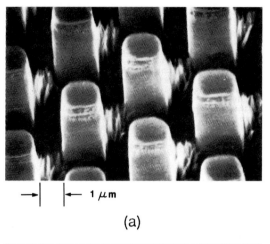

(a)

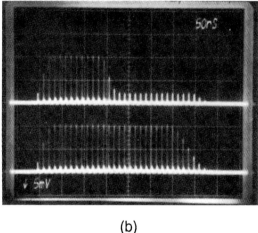

(b)

Fig. 5. (a) Array of GaAs – AlAs bistable optical microresonators with 2 μm typical element size. (b) Oscilloscope display of the pulse trains. From [McJ].

Thus, even the possibilities of binary digital optical triggers dedicated for more or less traditional computer architecture should not be deemed completely realized. The prospects for use of the methods of nonlinear optics and laser physics in optical computer technology seem to be much stronger. Modern nonlinear optics allows the development of fast amplitude, phase, and polarization triggers operating not only with discrete signals, but also with wave structures. In general, the potential of digital and analog optical computers is decided by advances in the development of efficient nonlinear materials, systems for generation and handling pico- and femtosecond laser pulses, and the methods of control of the longitudinal and transverse interactions of light waves in nonlinear media.

3 Strong Optical Nonlinearities.
Nonlinear Materials

In regard to the expansion of macroscopic polarization set up in the nonlinear material in powers of the field amplitude, optical computer technology is most interested in the nonlinear response leading to the term

$$\mathbf{P}_{nl}^{(3)} = \hat{\chi}^{(3)} \mathbf{EEE}, \tag{4}$$

with the susceptibility $\hat{\chi}^{(3)}$ being a tensor of fourth rank. This cubic nonlinearity gives rise to the nonlinear addend to the refractive index (1). In a medium described by (4), the harmonic wave field of frequency ω and amplitude A bring about the nonlinear polarization

$$\mathbf{P}_{nl}^{(3)}(\omega) = \hat{\chi}^{(3)} \mathbf{eee} A^2 A^*, \tag{5}$$

where $\mathbf{e}$ is the unit vector of polarization. The immediate implication is that the full refractive index of the medium with cubic response will be $n = n_0 + n_2 I$, with n_2 being proportional to $\hat{\chi}^{(3)}(\omega)$,

$$n_2 = (2\pi/n_0)^2 \hat{\chi}^{(3)}(\omega). \tag{6}$$

Cubic optical nonlinearity is virtually omnipresent – the cubic term (4) arises in gases, liquids, amorphous bodies, and crystals. This is explained by the fact that the cubic susceptibility $\chi_{ijkl}^{(3)}$ is also different from zero in media having a center inversion.

Physical mechanisms underlying the cubic response and, hence, the nonlinear refractive index are rather diverse. They are convenient to categorize with reference to important application parameters: the magnitude of the nonlinear coefficient n_2 and the relaxation time of nonlinear response, τ_{nl}. For the fastest ($\tau_{nl} \simeq 10^{-14}$ s) nonresonant, nonlinear response of optical electrons in transparent condensed medium, we have $\hat{\chi}^{(3)}(\omega) \simeq 10^{-14}$ csu and, respectively, $n_2 \simeq 10^{-13}$ cm^2/kW.

However, many other mechanisms exist that are not that fast, but can cause stronger nonlinearities in the refractive index. The list includes the resonant nonlinearities in semiconductors (electronic resonances in two-dimensional structures are of special interest; their nonlinear response is exploited in the bistable microresonators depicted in Fig. 5), the photorefractive effect in inorganic crystals, orientation of anisotropic molecules in the light field, and optical heating of the medium.

Experimental evidence related to different nonlinear materials is collated in Fig. 6.

The range of values of the nonlinear parameter n_2 currently available in nonlinear optics is seen to span *ten orders* of magnitude! Despite the different physical mechanisms, many data points cluster well around the lines $n_2 \sim \tau_{nl}$, a higher magnitude of nonlinear response being gained at the expense of longer switching time.

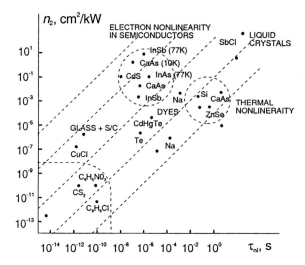

Fig. 6. Nonlinear parameter n_2 versus the nonlinear response time τ_{nl}.

In liquid crystals, the orientational mechanisms give rise to $n_2 > 0.1 \text{ cm}^2/\text{kW}$, which is quite appropriately termed *giant nonlinearity*. Strong nonlinearities of liquid crystals proved to be a convenient vehicle for modeling new types of optical triggers capable of handling wave structures in nonlinear analog computers.

4 Generation and Transformation of Femtosecond Light Pulses

Generation and transformation of femtosecond light pulses of a length of about 10^{-15} s whose envelope covers only a few periods of the light oscillations are some of the most remarkable achievements of laser physics over the last few years [AVC]. Research in femtosecond laser technology has secured a strong foothold for realization of the ultimate switching time of optical switching devices. An important contribution in this development has been fiber light guides – a virtually ideal medium for transmission and transformation of femtosecond pulses.

Figure 7 shows the schematic of an optical compressor designed to reduce the length of ultrashort light pulses.

We note that precisely this sort of system has been used to obtain ultimately short light pulses. As with optical bistability, the underlying principle of this system relies on the nonlinearity of the refractive index. A light pulse of the form

$$E = A(t) \exp i(\omega_0 t - k_0 z), \tag{7}$$

propagating in a medium with a nonlinear refractive index (1) experiences a phase self-modulation. Indeed, the total phase shift acquired by the pulse over

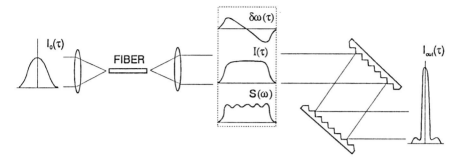

Fig. 7. Arrangement for compression of laser pulse duration with the aid of phase self-modulation in an optical fiber. Intensity profiles, frequency modulation $\delta\omega(\tau)$, and the spectrum $S(\omega)$ after the phase modulator are indicated for specific cross sections.

a distance z is

$$\varphi = kz = (\omega_0/c)(n_0 + n_2 I(\tau))z = k_0 z + k_0 n_2 I(\tau)z, \tag{8}$$

where τ is the time in the running coordinate system.

The time dependence of the intensity $I(\tau) = (c/4\pi)A^2(t)$ brings about the intensity-dependent nonlinear phase increment

$$\varphi_{nl}(\tau) = -k_0 n_2 I(\tau)z, \tag{9}$$

and so the frequency self-modulation of the pulse

$$\Delta\omega_{nl}(\tau) = \frac{\partial\varphi_{nl}}{\partial\tau} = -k_0 n_2 \frac{\partial I}{\partial\tau}z. \tag{10}$$

The resultant frequency spectrum of the pulse expands considerably. Launching the phase self-modulated pulse in an appropriate dispersive medium can cut the pulse length down to

$$\tau_{nl} \approx \frac{1}{\Delta\omega_{nl}} \approx \frac{\tau_0}{k_0 n_2 I_0 z}, \tag{11}$$

where τ_0 is the initial pulse length and I_0 is the peak intensity.

In media with $n_2 > 0$, the frequency of the self-modulated pulse increases from the leading edge to the trailing edge. Clearly, to compress the pulse, a medium with anomalous dispersion is needed. In the arrangement of Fig. 7, the role of such a medium is played by crossed diffraction gratings.

Phase self-modulation, or cross modulation if two or more pulses are propagating in the medium, is a fundamental nonlinear effect of the variety used in optical switches.

Generation of optical solitons has great importance for data transmission in optical computers. Solitons are stable pico- and femtosecond light pulses of invariable shape that propagate in nonlinear media at the balance between nonlinear compression and dispersion spreading. In optical fibers with $n_2 > 0$, solitons seem to be excited in the region of anomalous dispersion of the group velocity

[Mol]. Compression, cross modulation, and generation of optical solitons are the principal phenomena allowing one to tailor the advances in pico- and femtosecond laser technology for data control in optical computers (for more details, see [AVC, Mol]).

5 Control of Transverse Interactions in Nonlinear Optical Resonators: Generation, Hysteresis, and Interaction of Nonlinear Structures

Bistable optical systems outlined in the earlier sections (see also Figs. 2 through 5) are essentially the basic elements of digital optical computers.

Research in recent years indicates that nonlinear optics is capable of furnishing the basic elements for the hardware of neural network computers. This type of computer technology is being designed for irregular problem solving, pattern recognition, and artificial intelligence modeling. Several laboratories have employed for this purpose phase-conjugating systems and arrays of nonlinear switching elements in conjunction with holographic elements and light-field analyzers (see, e.g., [AMP]).

Another direction of optical modeling of a system of neurons interconnected in a sophisticated network is extending the idea of two-dimensional feedback (see [ALV, AkV]).[4] If one inserts in the feedback loop of a nonlinear ring resonator (of the type depicted in Figs. 4a and 11b), a field transformer to implement rotation and compression, the ordinary amplitude optical bistability and the associated temporal instabilities give way to a whole hierarchy of new nonlinear wave phenomena. This statement is illustrated in Fig. 8 by the photographs of typical nonlinear structures (from [AVI]).

These are rotating waves (optical reverberators), spiral waves that occur at large nonlinearity parameter $K = (2\pi/\lambda)n_2 IL$ due to the interaction of nonlinear structures, and random nonlinear fields representing optical turbulence. Thus, this list is an optical reproduction of the entire spectrum of phenomena of nonlinear wave dynamics, which are being extensively investigated in hydrodynamics, plasma physics, and biology, etc. [GaR].

Workers in the field of neural computers, pattern recognition, and associative memory are primarily interested in the possibility of control of the said structures and in their interactions. Some notable results have already been obtained in this way. As an example, Fig. 9 shows an experimental curve of the rotation velocity Ω of a reverberator with wave number N (number of petals over a period of 2π) as a function of initial rotation angle θ_0 [ILV].

For the nonlinearity relaxation time τ, this velocity is given by the expression

$$\Omega = \gamma K \frac{\tan N\theta_0}{\tau}. \tag{12}$$

At the plot, the characteristic points of transition are indicated in which the number of petals varies. The optical reverberator is basically a new type of wave

---

[4] Among more recent publications are [Vor, VRI].

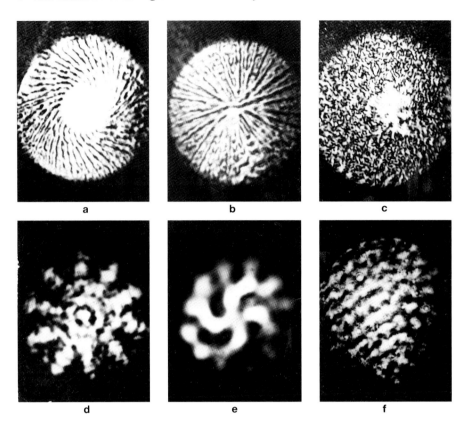

Fig. 8. Nonlinear structures in a resonator with 2-D feedback: (a, b, d, e) rotating waves (optical reverberators); (d) and (e) correspond to a lower diffusion coefficient; (c) developed optical turbulence; and (f) transition from rolls to optical turbulence (from [AVI]).

trigger that exhibits a pronounced hysteretic behavior. However, this time, unlike simple amplitude bistability, we speak about the hysteresis of structures rather than signals [ALV].

An illustration of this last thesis is given in Fig. 10.

The theoretical and experimental plots of Ω versus θ_0 constructed for both directions of θ_0 exhibit a clear-cut hysteresis. Good agreement between the experimental and theoretical data is noteworthy as an indication of the efficiency of the theoretical approach developed in [ALV, AkV].

The experimental and theoretical evidence given above gives us solid grounds to expect substantial modeling of neural structures with the aid of nonlinear resonators with 2-D feedback (Fig. 11) [AVI, Vor].

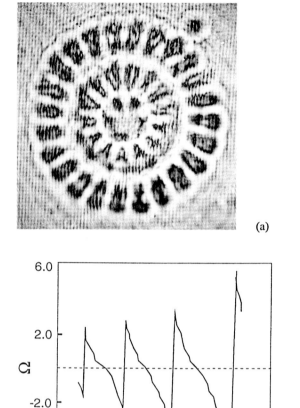

(a)

(b)

Fig. 9. Optical reverberator dynamics: (a) N-petaled structure, (b) plot of the rotation velocity versus the angle θ_0 of the initial field rotation in the feedback loop.

6 Conclusion. Nonlinear Optics and Molecular Electronics

Advances in the development of bistable optical components using semiconductor microresonators and nonlinear optical systems for transmission and transformation of ultrashort light pulses have stimulated a number of projects of digital all-optical computers (see, e.g., [McJ, Loh, OpB]). The question of whether such computers are competitive and have a certain field of application may be answered before long. On the other hand, it is clear that these examples do not exhaust the application of nonlinear optics to computer technology.

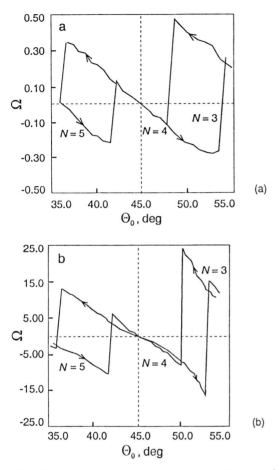

Fig. 10. Optical reverberator as a spatial trigger: (a) Theoretical and (b) experimental plots of angular velocity as a function of the initial rotation angle θ_0 of the field in the feedback loop reveal hysteresis cycles for various directions of θ_0.

Among the new directions of research, we mention the modeling of neural networks and the use of nonlinear optics in molecular electronics. Large organic molecules possess a very strong cubic response, therefore, the response of strongly delocalized electrons draws considerable attention [Duc]. The effect of electronic delocalization on the magnitude of $\hat{\chi}^{(3)}$ in organic molecules is presented in Fig. 12.

Strong additional growth ($10^4 - 10^5$ times) of the cubic susceptibility $\hat{\chi}^{(3)}$ can be achieved by carrying the molecule in an excited electronic state. Thus, integration of the methods of nonlinear optics with the well-established techniques of molecular electronics (see [Laz]) may open up new horizons in the development of new generation computers. It is safe to say that wide use of the methods of nonlinear optics in the interests of computer technology is in its early stage.

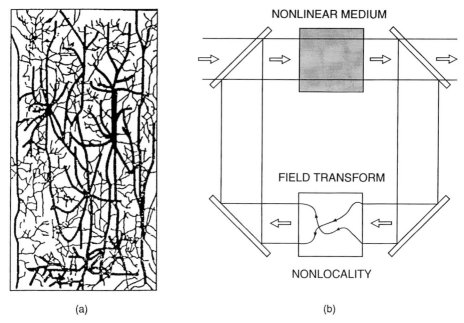

(a)   (b)

Fig. 11. Nonlinear optics modeling of neural networks (a) by nonlinear resonators with 2-D feedback (b).

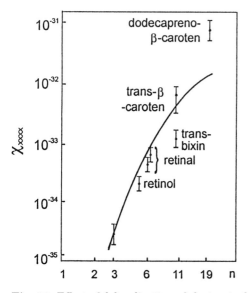

Fig. 12. Effect of delocalization of electronic density in organic molecules with n double bonds on the magnitude of cubic susceptibility $\hat{\chi}^{(3)}$. Experimental data points and an analytical curve. From Ducuing in *Nonlinear Spectroscopy*, eds. Bloombergen and Bologna, 1979.

References

[AMP] Abu-Mostafa, Y.S., Psaltis, D.: Optical Neuron Computers. Sci. Amer. **256** 1987, 88-95.

[AK1] Akhmanov, S.A., Khokhlov, R.V.: On Trigger Properties of Nonlinear Wave-Guiding Systems. Izv. Vyssh. Uchebn. Zaved. Radiofiz. **5** 1962, 742.

[AK2] Akhmanov, S.A., Khokhlov, R.V.: Optical Parametric Amplifiers and Oscillators. Usp. Fiz. Nauk **88** 1966, 493.

[ALV] Akhmanov, S.A., Larichev, A.V., Vorontsov, M.A.: Large Scale Transverse Interactions in Nonlinear Optical Systems with 2-D Feedback: Generation, Hysteresis, and Interaction of Wave Structures. In Coherence and Quantum Optics, **6** 1989, Plenum Press, New York.

[AkR] Akhmanov, S.A., Roshal, A.S.: Ternary Triggers Built around Parametric Oscillators. Izv. Vyssh. Uchebn. Zaved. Radiofiz. **5** 1962, 1017.

[AVC] Akhmanov, S.A., Vysloukh, V.A., Chirkin, A.S.: In Optics of Femtosecond Laser Pulses, American Institute of Physics, 1992; Nauka, Moscow, 1988 (in Russian).

[Duc] Ducuing, J.: Optical Nonlinearities of Conjugate 1-D Systems. In Nonlinear Spectroscopy. Ed. Bloembergen, N., Societa Italiana di Fisica, Bologna, 1979.

[GaR] Gaponov, A.V., Rabinovitch, M.I.: Autostructures: Chaotic Dynamics of Ensembles. In Nonlinear Waves. Nauka, Moscow, 1987 (in Russian).

[Gib] Gibbs, H.: Optical Bistability: Controlling Light with Light. Academic Press, Orlando, 1985.

[Got] Goto, E.: The Parametron, a Digital Computing Element which Utilizes Parametric Oscillations. Proc. IRE. **47** 1959, 1347.

[KeA] Keyes, R., Armstrong, J.A.: Thermal Limitation in Optical Logic. Appl. Opt. **8** 1969, 2549.

[Lan] Landauer, R.: A Personal View on Nonlinear Electro-Magnetic Wave Propagation. IBM Thomas Watson Research Center Memorandum, 1985.

[Loh] Lohmann, A.: What Classical Optics Can Do for the Digital Optical Computer. Appl. Opt. **25** 1986, 1543.

[Lug] Lugovoi, V.N.: On Stimulated Sum and Difference Mixing Oscillation in Optical Resonator. Zh. Eksp. Teor. Fiz. **56** 1969, 996.

[Man] Mandelshtam, L.S.: Collection of Works. Izd. Akad. Nauk SSSR **1-5**, Moscow.

[McC] McCall, S.L., Gibbs, H.M., Churchill, G.G., Venkateson, T.N.C.: Optical Transition and Bistability. Bull. Am. Phys. Soc. **20** 1975, 636.

[McJ] McCall, S.L., Jewell, J.: Monolithic Microresonator Arrays of Optical Switches. In Laser Optics of Condensed Media. Plenum Press, New York, 1986.

[Mol] Mollenauer, L.: Solitons in Optical Fibers. In Workshop on Structure, Coherence and Chaos in Dynamical Systems, MIDIT, 1986.

[Pap] Papaleksi, N.D.: Collection of Works. Izd. Akad. Nauk SSSR, Moscow, 1948.

[She] Shen, I.R.: The Principles of Nonlinear Optics. J. Wiley & Sons, New York, 1984.

[Ste] Sterzer, F.: Microwave Parametric Subharmonic Oscillators for Digital Computing. Proc. IRE **47** 1959, 1317.

[Szo] Szöke, A., Daneu, V., Goldhar, J., Kurnit, N.: Bistable Optical Element and Its Application. Appl. Phys. Lett. **15** 1969, 376.

[Wig] Wigington, R.L.: A New Concept in Computing. Proc. IRE **47** 1959, 516.

[Laz] Biomolecular Electronics and Self-Assembly Problems of Supermolecular Structure. Ed. Lazarev, P.I., Izd. Akad. Nauk SSSR, Moscow, 1987.

[OpB]   From Optical Bistability towards Optical Computing. Eds. Mandel, P., Smith, S.D., Wherrett, B.S., Elsevier Science Publishers, Amsterdam, 1987.

[AkV]   Akhmanov, S.A., Vorontsov, M.A., Ivanov, V.Yu., Larichev, A.V., Zheleznykh, N.I.: Controlling Transverse-Wave Interactions in Nonlinear Optics: Generation and Interaction of Spatiotemporal Structures. JOSA B **9** 1992, 78.

[VPS]   Vorontsov, M.A., Pruidze, D.V., and Shmalhauzen, V.I.: Spatial Bistability in Nonlinear Optical System with Optical Feedback. Izv. Vyssh. Uchebn. Zaved. Radiofiz. **25** Dec. 1988, 505.

[VZI]   Vorontsov, M.A., Zheleznykh, N.I., and Ivanov, V.Yu.: Transverse Interactions in 2-D Feedback on Non-Linear Optical Systems. Opt. and Quantum Electron. **22** 1990, 501.

[AVL]   Akhmanov, S.A., Vorontsov, M.A., and Larichev, A.V.: Dynamics of Non-Linear Rotating Light Waves: Hysteresis and Interaction of Wave Structures. Sov. J. Quantum Electron. **22** Apr. 1990, 325.

[Vor]   Vorontsov, M.A.: Problems of Large Neurodynamics System Modelling: Optical Synergetics and Neural Networks. Proc. Soc. Photo-Opt. Instrum. Eng. **1402** 1991, 116.

[VRI]   Vorontsov, M.A.: "Akhseals" as a New Class of Spatio-Temporal Light Field Instabilities. Sov. J. Quantum Electron. **20** 1993, 319.

[AVI]   Akhmanov, S.A., Vorontsov, M.A., Ivanov, V.Yu., Larichev, A.V., and Zheleznykh, N.I.: Controlling Transverse-Wave Interactions in Nonlinear Optics: Generation and Interaction of Spatiotemporal Structures. J. Opt. Soc. Am. B, 9 **1** Jan. 1992, 78.

[ILV]   Ivanov, V.Yu., Larichev, A.V., and Vorontsov, M.A.: 1-D Rotatory Waves in the Optical System with Nonlinear Large-Scale Field Interactions. Proc. Soc. Photo-Opt. Instrum. Eng. **1402** 1991, 145.

2 Optical Design Kit of Nonlinear Spatial Dynamics

E. V. Degtiarev[1,2] *and M. A. Vorontsov*[1,2,3]

[1] New Mexico State University, NM 88003, USA
[2] International Laser Center, Moscow State University 119899 Moscow, Russia
[3] U.S. Army Research Laboratory, White Sands Missile Range, N.M. 88002, USA

Introduction

Spatial and temporal complexity is one of the most interesting subjects of modern science. Just as understanding the composition of each individual building block does not automatically imply understanding the architecture of the building in general, the behavior of complex nonlinear systems consists of much more than the simple arithmetic sum of the component influences. It appears that the time of crushing complex systems into discrete "stones" and scattering these "stones" has ended, and we are turning at last to the creative work of collecting these elementary building blocks.

Here we show how to design a great variety (that is, a kaleidoscope) of spatial self-organization phenomena in nonlinear optics using a few rather simple building blocks – nonlinear optical systems with optical and electronic feedback circuits. We discuss "artificial complexity," or the complexity that arises not from natural phenomena and systems, but as a result of artificially designed interactions. Nevertheless, artificial complexity represents real physical processes that can and do occur in real optical systems. This work originally derives from the interdisciplinary fields of synergetics [Hak] and artificial neural network theory [Grs]. The traditional concepts of these two fields can be significantly expanded on when the features unique to optical systems are considered.

In Sect. 1, we discuss one-component nonlinear optical systems with 2-D feedback. These one-component optical systems, or synergetic optical blocks, serve as elementary building blocks for the design of nonlinear spatial dynamics. In Sect. 2, these systems are supplemented with an external 1-D electronic feedback circuit. This circuit is controlled by the total intensity of the optical feedback, thus introducing a new type of integral transverse interaction. Section 3 concerns the optical counterparts of two-component reaction-diffusion systems. We conclude by discussing how to create increasingly complex spatio-temporal dynamics using new types of local and long-range interactions.

1 Elementary Optical Synergetic Blocks

1.1 Characteristics of a Synergetic Block

The possibilities for extensive experimental investigation of complex synergetic systems are at present limited. For real-world phenomena, observation of and control over the processes under study usually involve the application of sophisticated experimental techniques. Accordingly, the majority of potential complexity remains beyond the scope of experimental investigation. Certainly one can refer to a computer, but as a rule, complex dynamics are too complex even for today's powerful computers. In this connection, the problem of devising reasonably universal "artificial" spatial and temporal complexity for modeling self-organization and chaotic behavior is of great importance.

A well-known source of spatio-temporal complexity in synergetics is the multicomponent reaction-diffusion-type system [NiP, VRY]:

$$\tau_1 \frac{\partial u_1}{\partial t} = D_1 \nabla_\perp^2 u_1 + f_1(u_1, \ldots, u_N, \boldsymbol{K}),$$

$$\vdots$$

$$\tau_N \frac{\partial u_N}{\partial t} = D_N \nabla_\perp^2 u_N + f_N(u_1, \ldots, u_N, \boldsymbol{K}), \tag{1}$$

where $\nabla_\perp^2$ is the Laplacian operator with respect to spatial variables; $u_n(\boldsymbol{r}, t)$ $(n = 1, \cdots, N)$ are functions that characterize the state of individual components; τ_n are relaxation times that determine the temporal scale of changes in the u_n variables; and f_n are nonlinear functions, among which are the so-called "N-like" functions. The diffusion coefficients D_n set the spatial scale of interactions with characteristic lengths $l_n = D_n^{1/2}$, and the control parameters $\boldsymbol{K}$ describe the system's excitation level.

For the case of $N = 2$, we arrive at a classical model for the two-component reaction-diffusion system [Hak]

$$\tau_u \frac{\partial u}{\partial t} = D_u \nabla_\perp^2 u + f(u, v, \boldsymbol{K}),$$

$$\tau_v \frac{\partial v}{\partial t} = D_v \nabla_\perp^2 v + g(u, v, \boldsymbol{K}). \tag{2}$$

The system (2) provides a fairly simple mathematical model for analyzing a great variety of self-organization phenomena in diverse nonequilibrium systems, including chemical reactions, biological systems, thermal conductivity, etc. By altering the parameters, many different patterns can be obtained, such as dissipative structures, traveling waves, and various auto-wave regimes [Krm].

We propose building optical analogues of multicomponent reaction-diffusion systems from elementary optical blocks. What features would be required for such a synergetic optical block?

(1) The dynamics of each elementary block must be described by the reaction-diffusion-type equation [1][1]

$$\tau \frac{\partial u}{\partial t} = D\nabla_\perp^2 u + f(u, \boldsymbol{K}).$$ (3)

with N-like nonlinear function $f(u, \boldsymbol{K})$.
(2) Optical building blocks should be structured so as to permit various combinations that form multicomponent systems of the reaction-diffusion type (1).
(3) The dynamic variables describing the current state of the system must be easily observed and measured.
(4) The primary parameters $\boldsymbol{K}$ of the individual blocks and their interactions must be controllable.

Optical systems with these properties will have all the features necessary to be the elementary building blocks of complex multicomponent systems. Taken together these elemental blocks constitute a "design kit" for complex spatio-temporal dynamics. Although the idea of such a design kit may seem rather unusual, it can be realized using nonlinear optical systems with 2-D feedback. The simplest nonlinear systems with 2-D feedback, representing our most elementary synergetic blocks, have been investigated in a number of works [VKN, FiW, VKS].

1.2 Optical Synergetic Block Based on LCLV

Spatio-temporal light modulators form the basis of an optical synergetic block [VKN, FiW]. Experiments have been performed using both liquid-crystal light valves (LCLVs) and electro-optical modulators [Vor, AkV]. Consider the typical structure of a spatio-temporal modulator containing liquid crystal (Fig. 1) [Vas]. The LCLV is a multilayered structure sandwiched between two glass plates having transparent conductive electrodes. A photoconductor layer and dielectric mirror are deposited on one glass plate. An opaque layer between the photoconductor and mirror prevents the input light from affecting the photoconductor. Between the photoconductor and the second plate, there is a thin ($\sim 10\mu$) layer of nematic liquid crystal (LQ). A special layer on the glass plates causes the axes for the liquid crystal molecules (the LQ director) to be oriented parallel to the glass plates. The liquid crystal cross section shown in Fig. 1 depicts a typical optical crystal with two axes. Voltage is applied to the LCLV electrodes. The intensity distribution of the controlling light illuminating the LCLV photosensitive layer results in a spatial modulation of the photoconductor resistance. This change in local resistance, in turn, affects the spatial modulation of the voltage applied to the LQ layer. The voltage variation gives rise to a spatial modulation of the liquid crystal layer's refractive index $\Delta n(\boldsymbol{r}) = n_\parallel(\boldsymbol{r}) - n_\perp$. Here, $n_\parallel$ is the refractive index component corresponding to the director of the LQ molecules (that is, the liquid crystal's optical axis). The component $n_\perp$ corresponds to the

---

[1] Equation (3) is also referred to as the FKPP (Fisher-Kolmogorov-Petrovskii-Piskunov) equation [Fis, KPP].

orthogonal direction. Note that the controlling light intensity leaves $n_\perp$ unaffected. In fact, $n_\parallel$ and $n_\perp$ are the extraordinary and ordinary components of the liquid crystal's refractive index.

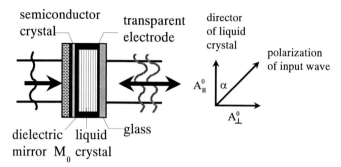

Fig. 1. The liquid crystal light valve.

Assume that the input wave is linearly polarized and that the polarization direction forms an angle α with the director of the liquid crystal. Denote the complex amplitude of the input wave by A^0. The two orthogonal projections of this field are designated as $A_\parallel^0 = A^0 \cos(\alpha)$ and $A_\perp^0 = A^0 \sin(\alpha)$ (parallel and perpendicular to the LQ optical axis, respectively). On passing through the LQ layer and reflecting off the mirror M_0, the input field obtains the additional phase modulation

$$A_\perp' = A_\perp^0 \exp(i\phi_0), \quad A_\parallel' = A_\parallel^0 \exp(iu + i\phi_0), \tag{4}$$

where $\phi_0 = 2lkn_\perp$ and $u(\boldsymbol{r}, t) = 2lk\Delta n(\boldsymbol{r}, t)$, l is the thickness of the layer, and $k = 2\pi/\lambda$ is the wave number. Note that the phase u varies due to the action of the controlling light on the LCLV photosensitive layer, but ϕ_0 remains unchanged.

Figure 2 shows the typical modulation characteristic of an LCLV [VoL]. The continuous curve corresponds to an approximation of the experimental data using the function $u = f_m(I_u) = p\,\text{th}(bI_u + c)$, where I_u is an external controlling light intensity. The parameters p, b, and c depend on the type of LCLV and its mode of operation. In the interval $0.05 < I_u < 0.30 \text{ mW/cm}^2$, the modulation characteristic is linear. In this intensity range the LCLV acts as a thin slice of a Kerr medium with an extremely large equivalent nonlinearity parameter $n_2 = 0.12 \text{ cm}^2/\text{mW}$ [VoL]. By choosing the amplitude and frequency of the voltage applied to the LCLV, one can change the value of the nonlinearity.

A simple mathematical model for the LCLV that considers carrier diffusion in the photoconductor and the finite response time of the liquid crystal can be written in the form [Vor, AkV]

$$\tau \frac{\partial u}{\partial t} + u = D\nabla_\perp^2 u + f_m(I_u), \tag{5}$$

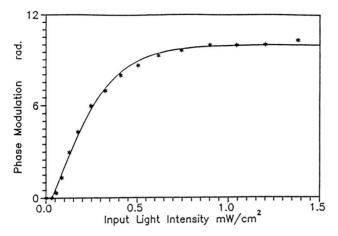

Fig. 2. Modulation characteristic of an LCLV. Experimental data approximated by a continuous curve.

where D is the diffusion coefficient. We neglect longitudinal diffusion because of the small thickness of the layer ($\sim 10\mu$).

Equation (5) differs from the FKPP equation (3) for an elementary synergetic block in that the nonlinear function f does not depend on the current state of the system (on the variable u). To obtain an equation of the FKPP type, the wave reflected from the LQLV should be used as the controlling field. This is possible in a system with 2-D feedback (Figs. 3 and 4).

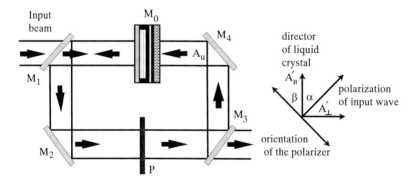

Fig. 3. Optical feedback system using a polarizer to transform spatial phase modulation into the corresponding intensity distribution.

In Fig. 3, the feedback circuit is formed by the mirrors $M_1 - M_4$ and the polarizer P, which is at an angle β with the director of the liquid crystal molecules. The complex amplitude of the controlling light A_u is determined by projections

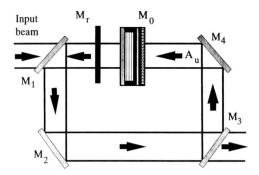

Fig. 4. Interferometric scheme of the optical synergetic block.

of the wave components $A'_\perp$ and $A'_\parallel$ on the direction of the polarizer P (see Fig. 3)

$$A_u = A^0_\parallel \cos(\beta) \exp(iu + i\phi_0) - A^0_\perp \sin(\beta) \exp(i\phi_0),$$

or in accordance with (4)

$$I_u = \mid A_u \mid^2 = \kappa I^0[1 + \gamma \cos(u)], \tag{6}$$

$$\kappa = \cos^2(\alpha)\cos^2(\beta) + \sin^2(\alpha)\sin^2(\beta), \quad \gamma = -2\cos(\alpha)\cos(\beta)\sin(\alpha)\sin(\beta)/\kappa.$$

In fact, the controlling light intensity (6) depends on the phase modulation u, just as does the intensity distribution in the interference pattern in the scheme shown in Fig. 4.

1.3 Main Mathematical Models

The system with optical feedback discussed above can be described by the equation

$$\tau \frac{\partial u}{\partial t} + u = D\nabla^2_\perp u + f_{\mathrm{m}}(\kappa I^0[1 + \gamma \cos(u)]), \tag{7}$$

where the function f_{m} describes the modulation characteristic of the LCLV. For simplicity, here we restrict consideration to the interval where f_{m} is linear. This allows use of equation (7) for pure optical Kerr-type nonlinearity.

Taking into account the optional illumination of the LCLV photosensitive layer by an external light with intensity distribution I_{ext}, we obtain

$$\tau \frac{\partial u}{\partial t} + u = D\nabla^2_\perp u + K[1 + \gamma \cos(u + \varphi)] + \chi I_{\mathrm{ext}}. \tag{8}$$

The nonlinear interaction is specified by the parameter $K = \kappa \chi I^0$, where χ is the slope for the modulation characteristic function (an analogue to the nonlinear refractive index coefficient for Kerr-type media). The additional phase shift φ entering (8) can be varied by altering the voltage amplitude a_φ applied to the LCLV.

Equation (8) also describes the dynamics in a nonlinear interferometer with 2-D feedback (Fig. 4) [VKS, VoL]. In comparison with the configuration discussed above (Fig. 3), the interferometer in Fig. 4 uses an additional reference mirror M_r to transform the nonlinear phase modulation $u(r,t)$ into the corresponding intensity distribution I_u. In this case, there is no need to use a polarizer. The LQ optical axis should be oriented in the direction of the input wave polarization.

We see that both optical systems meet the requirements for a synergetic optical block:

(1) Equation (9) is a typical FKPP equation with an N-like type of nonlinear function.

(2) Due to external optical and electronic inputs (the light intensity $I_{ext}(r,t)$ and voltage amplitude a_φ, see Fig. 5) individual blocks can be combined to form multicomponent systems.

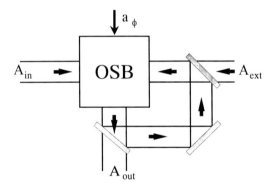

Fig. 5. Schematic of the optical synergetic block.

(3) The output signal for both optical systems is the intensity distribution in the optical feedback, or the pattern

$$I_{out}(r,t) = \mid A_{out}(r,t) \mid^2 = \eta[1 + \gamma \cos(u + \varphi)], \tag{9}$$

which can be easily measured and observed. The characteristic response time of the LQLV is $10^{-1} - 10^{-3}$s, which significantly simplifies detection of dynamic processes.

(4) It is simple to experimentally vary the following parameters: K (input intensity in the feedback loop), φ (phase shift between the object and reference waves in the interferometer, and/or the applied voltage amplitude a_φ), and I_{ext} (external light intensity). It is also possible to vary the diffusion coefficient D and relaxation time τ, for example, by changing the applied voltage frequency or construction of the LCLV [Vas].

1.4 Optical Multistability and Switching Waves

Consider the primary features of the dynamic behavior of the model (8). Spatially uniform steady states satisfy the equation

$$f(u) = -u + K[1 + \gamma \cos(u + \varphi)] = 0, \tag{10}$$

where $I_{\text{ext}} = 0$. For steady states, according to (10) the nonlinear phase modulation u is proportional to the feedback intensity $I_{\text{out}} = \eta[1 + \gamma \cos(u + \varphi)]$. Because of the cosine nonlinearity, (10) may have multiple solutions for sufficiently large K. Here we consider the simplest case, when $f(u) = 0$ has only three roots $u_1 < u_2 < u_3$ (see Fig. 6). The steady states u_1 and u_3 are asymp-

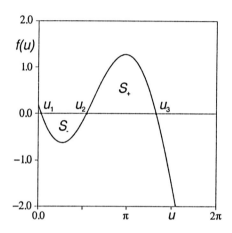

Fig. 6. The function $f(u) = -u + K[1 + \gamma \cos(u + \varphi)])$ for $\varphi = 2.7$, $K = 2.3$. $\varphi_s = \frac{3\pi}{2} - K \simeq 2.41$.

totically stable, while u_2 is unstable with respect to plane-wave perturbation. The system will correspondingly exhibit optical bistability [Gbs, VZI]. When two stable states coexist within the radial extent of the beam, they are separated by narrow $(l \sim \sqrt{D})$ transition layers that, in general, move, also called switching waves.

Switching waves, also known as fronts, for a one-dimensional $(u = u(x,t))$ bistable reaction-diffusion-type system (8) theoretically can be described as follows. A front moving at a constant velocity V is a solution to (8) that depends only on the traveling coordinate $\xi = x - Vt$, and asymptotically connects two steady states: $u(\xi) \rightarrow u_1$ for $\xi \rightarrow -\infty$, and $u(\xi) \rightarrow u_3$ for $\xi \rightarrow +\infty$, representing the boundary conditions. For the one-dimensional case of (8), by introducing the new independent variable ξ, we obtain

$$Du'' + V\tau u' - u + K[1 + \gamma \cos(u + \varphi)] = 0, \tag{11}$$

where the prime denotes differentiation with respect to ξ. For this specific choice of boundary conditions, a front with positive velocity produces switching from the upper state u_3 to the lower state u_1.

The nonlinear eigenvalue problem (11) is analytically insolvable for an arbitrary function $f(u)$. However, in the next section, we need to know only the sign of V, which determines the direction of propagation for the switching wave. This qualitative information can be derived from fairly simple geometric arguments.

Multiplying (11) by u' and integrating over ξ from $-\infty$ to $+\infty$, we obtain

$$V\tau \int_{-\infty}^{+\infty} (u')^2 d\xi + \int_{u_1}^{u_3} \{-u + K[1 + \gamma \cos(u + \varphi)]\} du = 0. \qquad (12)$$

The second term on the left side of (12) can be represented as

$$\int_{u_1}^{u_3} f(u) du = \int_{u_1}^{u_3} \{-u + K[1 + \gamma \cos(u + \varphi)]\} du = -S_- + S_+, \qquad (13)$$

where

$$S_- = \int_{u_2}^{u_1} f(u) du \text{ and } S_+ = \int_{u_2}^{u_3} f(u) du$$

denote the positive and negative contributions to the integral, respectively (see also Fig. 6). With (12) and (13), we have

$$\text{sign}(V) = \text{sign}(S_- - S_+). \qquad (14)$$

From (14) it follows that a switching wave has zero velocity (a stationary boundary) for $S_+ = S_-$, or equivalently

$$-\int_0^{u_2 - u_1} f(u_2 - \xi) d\xi = \int_0^{u_3 - u_2} f(u_2 + \xi) d\xi. \qquad (15)$$

The equality (15) is fulfilled identically if the function $f(u)$ satisfies

$$f(u_2 + \xi) = -f(u_2 - \xi) \text{ for all } \xi. \qquad (16)$$

To prove this, it suffices to show that (16) implies $u_3 - u_2 = u_2 - u_1$. Assuming (16) holds, let $\xi = u_2 = u_1$, so that

$$f(u_2 + (u_2 - u_1)) = -f(u_2 - (u_2 - u_1)) = -f(u_1) = 0. \qquad (17)$$

Because we assumed that only three roots of the equation $f(u) = 0$ existed, from (17) it follows that $u_3 = 2u_2 - u_1$. Thus, the limits of integration in (15) coincide.

Using the explicit form of the function $f(u)$, the conditions when (15) occurs can be derived. After algebraic manipulation, (16) yields

$$(u_2 - K) = K\gamma \cos(u_2 + \varphi) \cos(\xi). \qquad (18)$$

Clearly, (18) is independent of ξ if

$$u_2 + \varphi = (2k + 1)\frac{\pi}{2}, \quad k = 0, \pm 1, \pm 2 \ldots,$$

and $u_2 = K$ is required to satisfy (18). Thus the equality $S_+ = S_-$ and, hence, $V = 0$ holds, provided the parameters φ and K are related by the equation

$$\varphi = (2k + 1)\frac{\pi}{2} - K, \ k = 0, \pm 1, \pm 2 \ldots \tag{19}$$

To avoid ambiguity in the definition of φ, note that the function $f(u)$ is invariant under the transformation $\varphi \to \varphi + 2\pi m$, where m is an integer, so φ can be restricted to the interval $[0, 2\pi]$. Furthermore, as bifurcation analysis of the equation $f(u) = 0$ shows, assuming that the function $f(u)$ has three zeros holds only for those values of φ corresponding to odd k's in (19). For $K < \frac{3\pi}{2}$, the value of φ satisfying these conditions is $\varphi_s = \frac{3\pi}{2} - K$.

From Fig. 6, $S_+ > S_-$ for $\varphi > \varphi_s$. According to (14), velocity is negative and propagation of the front induces switching from the lower state to the upper state. This demonstrates that the dynamics of switching waves can be governed by varying the external parameters φ and K.

2 Integral Transverse Interactions

2.1 The Synergetic Optical Block with an Electronic Feedback Circuit

Switching waves represent a typical dynamic process in a nonlinear bistable system, where this bistable system is an elementary optical synergetic block. The presence of the external inputs $I_{\text{ext}}(r)$ and $\varphi(t)$ in the optical block allows the design of complex and interesting nonlinear dynamics, even for the case of an individual block.

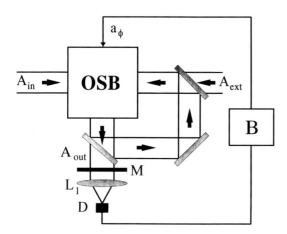

Fig. 7. Synergetic optical block with an electronic feedback circuit.

Consider the optical arrangement shown in Fig. 7. Part of the output wave for the optical synergetic block is directed into the opto-electronic circuit. This circuit consists of the optical portion (mask M and lens L_1), the registration scheme (photodetector D in the focal plane of lens L_1, and the electronic block B). By using the electronic block B, an external control signal $a_\varphi(t)$ for the LCLV can be produced. This completes an additional opto-electronic feedback circuit.

For simplicity, assume that the electronic block is an amplifier with input v and output a_φ voltages related by the function $a_\varphi = G(v)$, and that the phase shift φ is proportional to the controlling voltage $\varphi = \mu a_\varphi$. Denote the input resistance and capacitance of the amplifier by R and C, respectively. Evolution of the output signal applied to the LCLV is described by

$$C\dot{\varphi} = -\varphi/R + \mu G(v), \tag{20}$$

where $v = \beta \int M(\boldsymbol{r}) I_{\mathrm{out}}(\boldsymbol{r}, t) d^2 \boldsymbol{r}$, and β is a coefficient.

Using (8), (9), and (20), the mathematical model for the optical synergetic block with electronic feedback circuit can be written

$$\tau \frac{\partial u}{\partial t} + u = D\nabla_\perp^2 u + K[1 + \gamma \cos(u + \varphi)], \tag{21}$$

$$\tau_\varphi \dot{\varphi} + \varphi = \varphi_0 + R\mu G\{\beta \int M(\boldsymbol{r})[1 + \gamma \cos(u + \varphi)] d^2 \boldsymbol{r}\}. \tag{22}$$

In (21) the additional voltage φ_0 applied to the LCLV and the electronic circuit response time $\tau_\varphi = RC$ are accounted for.

There are additional opportunities for controlling spatio-temporal dynamics in the model given in (21). A mask $M(\boldsymbol{r})$ can be placed in the focal plane of lens L_1. In fact, by choosing various masks, the integral term in (21) is then dependent on different spatial frequencies of the optical field.

If the response function is $G(v) = -v$ and $M(\boldsymbol{r}) = 1$, the equation for φ takes the form

$$\tau_\varphi \dot{\varphi} + \varphi = \varphi_0 - \frac{\alpha}{S} \int_S K\{1 + \gamma \cos[u(\boldsymbol{r}, t) + \varphi(t)]\} d^2 \boldsymbol{r}, \tag{23}$$

where S denotes beam aperture and $\alpha = \mu R \beta S$ is an external parameter.

Dynamic behavior of the system (21) and (23) is best illustrated by a simple example. Set $\varphi_0 = \pi, \gamma = 1$, and let α vary near unity. The initial state of this system is then characterized by a unique spatially uniform steady-state solution $(\bar{u}, \bar{\varphi}) = (0, \pi)$. Consider evolution of a spatially localized perturbation of this trivial state. Figure 8 shows spatial distribution of the nonlinear phase modulation u at successive times. Due to the bistability of u, this perturbation causes local switching to the upper steady state u_3, which corresponds to a high-intensity value. The boundaries of this high-intensity domain rapidly form into steep fronts. Because $\varphi(0) = \pi > \varphi_s$, the switching waves start to move in opposite directions, expanding the bright area. According to (23), the consequent growth of total feedback intensity causes a decrease in φ. This process continues until φ reaches the value of φ_s corresponding to zero velocity of the switching wave. The pattern that eventually arises is a bright spot against a dark background.

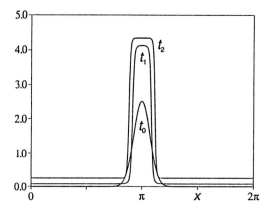

Fig. 8. Evolution of a localized initial perturbation. Numerical simulation in one transverse dimension x.

To estimate the size of the bright area, represent the true pattern by a stepwise function. Total feedback intensity is then given by

$$I_f = S_h u_3 + (S - S_h)u_1, \tag{24}$$

where S_h is the size of the high-intensity domain. Substituting (24) in (23), the relationship between the size of the bright spot and the system parameters is obtained

$$\frac{\alpha}{S}(S_h u_3 + (S - S_h)u_1) = \varphi_0 - \varphi_s. \tag{25}$$

The position of the bright spot that results from the transient process is determined by the initial conditions as given in the function $u_0(\boldsymbol{r}) = u(\boldsymbol{r}, 0)$. For a unimodal function $u_0(\boldsymbol{r})$, the bright spot forms in the vicinity of the extremum. Initial perturbation with several local extrema involves far more interesting dynamics. At the early stage of the transient process, bright spots start to form in the vicinity of each local extremum. However, the integral term in (23) invokes competition among these structures. The parameter α can be chosen to ensure that only the bright spot corresponding to the global extremum of the function $u_0(\boldsymbol{r})$ eventually "survives." The dynamics of this fairly simple system has certain features of WTA (winner takes all) competition [Kob, Grs]. In the two transverse dimensions, similar arguments apply for analysis of radially symmetric patterns.

3 Optical Counterparts of Two-Component Reaction-Diffusion Systems

The optical design kit concept suggests that complex spatio-temporal dynamics are the result of assembling elementary constituent parts, in this case, the optical synergetic blocks previously discussed. However, the basic elementary part by itself has a high level of intrinsic complexity. For simplicity, we begin by combining only two optical blocks.

The first question is, "How are interactions between individual subsystems introduced"? One possibility is represented in Fig. 9. Two 2-D optical feedback

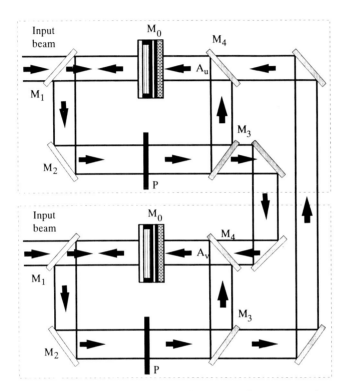

Fig. 9. General configuration of an optical system with two elementary synergetic blocks.

loops from the individual optical blocks are linked, so that the controlling light illuminating the photosensitive layer for the two LCLVs depends on the feedback intensity distributions for both subsystems. Let $u(\boldsymbol{r}, t)$ and $v(\boldsymbol{r}, t)$ denote nonlinear phase modulation of light reflected from the LCLV in the first and second blocks, respectively. Then the dynamics of u and v are governed by

$$\tau_u \frac{\partial u}{\partial t} + u = D_u \nabla_\perp^2 u + I_u,$$

$$\tau_v \frac{\partial v}{\partial t} + v = D_v \nabla_\perp^2 v + I_v, \tag{26}$$

where I_u and I_v are the intensity distributions of light in the plane of the photosensitive layer of each LCLV. According to Fig. 9,

$$I_u = K_{uu}[1 + \gamma \cos(u + \varphi_u)] + K_{uv}[1 + \gamma \cos(v + \varphi_v)],$$

$$I_v = K_{vu}[1 + \gamma \cos(u + \varphi_u)] + K_{vv}[1 + \gamma \cos(v + \varphi_v)], \tag{27}$$

where the coefficients K_{uu}, K_{vv}, K_{uv}, and K_{vu} are determined by the input light intensities in the subsystems and by the reflective coefficients of the mirrors and beam splitters in the optical feedback loops.

Nontrivial dynamics appear in this interconnected system, even if only one of the "self-action" terms in (27) is nonzero (K_{vv}, for example). This system is described by

$$\tau_u \frac{\partial u}{\partial t} = D_u \nabla_\perp^2 u + f(u, v, \mathbf{K}),$$
$$\tau_v \frac{\partial v}{\partial t} = D_v \nabla_\perp^2 v + g(u, v, \mathbf{K}). \tag{28}$$

The functions on the right side of (28) are given by

$$f(u, v, \mathbf{K}) = -u + K_{uv}[1 + \gamma \cos(v + \varphi_v)],$$
$$g(u, v, \mathbf{K}) = -v + K_{vu}[1 + \gamma \cos(u + \varphi_u)] + K_{vv}[1 + \gamma \cos(v + \varphi_v)]. \tag{29}$$

Figure 10 shows the corresponding optical arrangement. The feedback loop of one of the optical blocks is open, and as a result, the corresponding LCLV is illuminated only by light from the other optical block's feedback loop.

Equation (28) belongs to a wide class of two-component reaction-diffusion models (2) that have multiple applications in various fields of science. Despite many works devoted to the dynamics of two-component reaction-diffusion systems, complete understanding of their behavior is still lacking. Exact solutions to (28) cannot be obtained, except for a few special cases. However, several useful techniques of qualitative analysis have been developed. The following discussion is based mainly on such qualitative considerations. We hope this lack of mathematical rigor is admissible, as our main goal here is to review a wide range of patterns and instabilities in two-component optical systems, rather than to discuss any particular phenomenon at length.

As a first rough approximation, ignore the diffusion terms in (28) and assume that all spatial inhomogeneities can be disregarded. In (28), let $D_u = D_v = 0$, so that instead of (28), we obtain the dynamic system

$$\tau_u \frac{du}{dt} = -u + K_{uv}[1 + \gamma \cos(v + \varphi_v)],$$
$$\tau_v \frac{dv}{dt} = -v + K_{vu}[1 + \gamma \cos(u + \varphi_u)] + K_{vv}[1 + \gamma \cos(v + \varphi_v)]. \tag{30}$$

The original spatially extended system (29) can be regarded as a continuum of diffusively coupled individual elements with the dynamics governed by (30). Equation (30) is a significant improvement over the original problem, because there is an elaborate theory for 2-D dynamic systems [And]. However, (30) offers little information about what to expect in the spatially extended case.

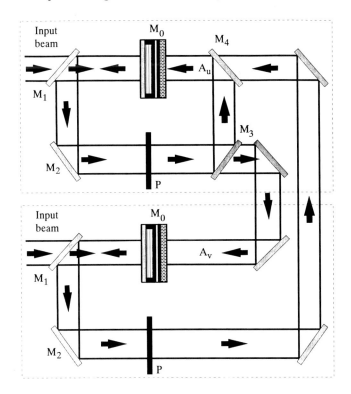

Fig. 10. Optical arrangement of the two-component system.

Evolution of (30) gives rise to solution curves or trajectories in the (u, v) plane. Eliminating explicit time dependence from (30) yields the equation for the solution curves

$$\frac{dv}{du} = \frac{\tau_u}{\tau_v} \frac{g(u, v, \boldsymbol{K})}{f(u, v, \boldsymbol{K})},$$ (31)

where f and g are defined by (29). Because nonlinearity (31) is still an analytically intractable problem, numerical simulation or a qualitative method must be used. The method of isoclines provides a useful technique for global approximation of trajectories. Curves, where the slope of the trajectories $dv/du = c$ are constant, are identified. Only the so-called nullclines, or isoclines, where dv/du is either zero or infinite, are considered. The nullcline for the function f is defined as the curve $u = u(v)$ given implicitly by $f(u, v, \boldsymbol{K}) = 0$. Similarly, the nullcline $v = v(u)$ is determined by $g(u, v, \boldsymbol{K}) = 0$. Intersection points for the nullclines give the equilibria of (30) and spatially uniform steady states of (28), respectively.

Referring to Fig. 11, which shows the nullclines for the system (30), notice the following:
(i) the functions $u = u(v)$ and $v = v(u)$ are both 2π-periodic
(ii) v is a multi-valued function of u

E.V. Degtiarev and M.A. Vorontsov

(iii) the intersection point of the nullclines changes position as the parameters are varied

Significantly, from Fig. 11a to b, the external parameters were varied only slightly, while the resulting change in dynamic behavior was dramatic. The next section shows that the assumption concerning the stability of spatially uniform states can be derived from analysis of the location where the nullclines intersect. It is easy to show that for steady states $(\bar{u}, \bar{v})$ located on the middle branch of the curve $v(u)$, such as the one in Fig. 11a,

$$g'_v(\bar{u}, \bar{v}) > 0.$$

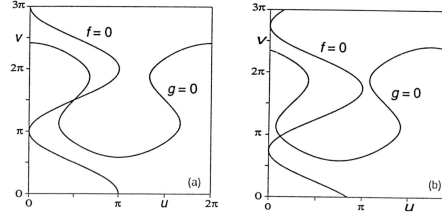

Fig. 11. Nullclines for the functions f, g. $K_{vv} = 2.5, K_{vu} = \frac{3\pi}{2} - K_{vv}, K_{uv} = \frac{\pi}{2}$. (a) $\varphi_u = \varphi_v = 0$. (b) $\varphi_u = \varphi_v = 0.8$.

This property of the function g is responsible for the instabilities associated with the spontaneous disintegration of spatial and temporal symmetry discussed in the next section. Due to the periodic character of the nullclines several intersection points can be obtained. This is a rather unusual situation for the classical theory of two-component systems, which traditionally operates using cubic-type nonlinear functions.

3.1 Linear Stability Analysis and Bifurcation of Uniform States

We will examine the stability of a spatially uniform steady state $(u(r,t), v(r,t)) = \bar{u}, \bar{v}$ with respect to small perturbations. We use a solution of the form

$$\begin{pmatrix} u(r,t) \\ v(r,t) \end{pmatrix} = \begin{pmatrix} \bar{u} \\ \bar{v} \end{pmatrix} + \begin{pmatrix} \delta u(r,t) \\ \delta v(r,t) \end{pmatrix} \tag{32}$$

in (28), keeping only the first-order terms with respect to δu and δv in the Taylor series expansion of the nonlinear functions. The resulting linear stability problem is

$$\begin{pmatrix} -\tau_u \frac{\partial}{\partial t} + D_u \nabla_\perp^2 + f_u' & f_g' \\ g_u' & -\tau_v \frac{\partial}{\partial t} + D_v \nabla_\perp^2 + g_v' \end{pmatrix} \begin{pmatrix} \delta u(\boldsymbol{r}, t) \\ \delta v(\boldsymbol{r}, t) \end{pmatrix} = 0, \qquad (33)$$

where primed quantities denote the first partial derivatives at the point $(\bar{u}, \bar{v})$.

In (33), spatial and temporal variables separate, so the solution should be of the form

$$\begin{pmatrix} \delta u \\ \delta v \end{pmatrix} = \begin{pmatrix} 1 \\ a \end{pmatrix} \exp(\sigma t)\psi_k(\boldsymbol{r}), \qquad (34)$$

where a is a constant, $\sigma = \nu + \imath\omega$, and ν and ω are the linear growth rate and frequency, respectively. The function $\psi_k(\boldsymbol{r})$ is compatible with boundary conditions and satisfies the equation

$$\nabla_\perp^2 \psi_k = -k^2 \psi_k,$$

where the spectrum of spatial frequencies k can be either discrete or continuous. In the following analysis boundary effects are ignored by assuming that the spatial extent of the system is infinite. Then $\psi_k(\boldsymbol{r})$ is simply a spatial Fourier component with wavevector $\boldsymbol{k}$: $\psi_k(\boldsymbol{r}) = e^{\imath \boldsymbol{k} \boldsymbol{r}}$.

To reduce the number of independent parameters, set $\gamma = 1, \varphi_u = \varphi_v = 0, K_{uv} = \frac{\pi}{2}$, and $K_{vu} = \frac{3\pi}{2} - K_{vv}$, assuming that $K_{vv} \equiv K$ is the parameter that varies. These restrictions are not physically important and are introduced here for convenience. Using these new parameters, it is easy to find the spatially uniform steady-state solution $(\bar{u}, \bar{v}) = (\frac{\pi}{2}, \frac{3\pi}{2})$, which may become unstable as K is increased. The partial derivatives entering (33) are the elements of the Jacobian matrix $[\partial(f, g)/\partial(u, v)]$ at the point $(\bar{u}, \bar{v})$

$$\begin{pmatrix} f_u' & f_v' \\ g_u' & g_v' \end{pmatrix} = \begin{pmatrix} -1 & \frac{\pi}{2} \\ K - \frac{3\pi}{2} & -1 + K \end{pmatrix}. \qquad (35)$$

However, it is more convenient to perform the calculations for the general case and use (35) only at the final stage.

Substituting (34) in (33) leads to a linear algebraic problem that ultimately yields the dispersion relation

$$\tau_u \tau_v \sigma^2 + \alpha(k)\sigma + \beta(k) = 0, \qquad (36)$$

where

$$\begin{aligned} \alpha(\boldsymbol{k}) &= \tau_u(k^2 D_v - g_v') + \tau_v(k^2 D_u - f_u'), \\ \beta(\boldsymbol{k}) &= (k^2 D_u - f_u')(k^2 D_v - g_v') - f_v' g_u'. \end{aligned} \qquad (37)$$

The roots of (36) are

$$\sigma_{1,2} = \frac{1}{2\tau_u \tau_v} \left(-\alpha \pm \sqrt{\alpha^2 - 4\tau_u \tau_v \beta} \right). \qquad (38)$$

An instability develops when the real part of σ increases through zero for a certain critical wavevector k_c. At the threshold $\nu = 0$, while the frequency ω may be either zero (for $\alpha^2 - 4\tau_u \tau_v \beta \geq 0$) or finite (for $\alpha^2 - 4\tau_u \tau_v \beta < 0$). The first case ($\omega = 0$) is expected to give rise to static spatially periodic patterns, also referred to as dissipative structures. This instability was first studied by Turing, who proved its relevance to the process of morphogenesis [Tur]. The occurrence of instability at finite ω is called a Hopf bifurcation [Mar] and is associated with oscillatory behavior.

In the case of a Hopf bifurcation, the real part of the complex conjugate roots $\sigma_{1,2} = \nu \pm \imath\omega$ becomes positive for

$$\alpha(k) = \tau_u(k^2 D_v - g_v') + \tau_v(k^2 D_u - f_u') < 0. \tag{39}$$

The quantity α as a function of k has a global minimum at $k = 0$ independently of the other parameters, and consequently, the most unstable mode has a wavevector of zero. In accordance with (38), at the threshold ($\alpha(0) = 0$), the oscillation frequency ω_0 is given by

$$\omega_0 = (\tau_u\tau_v)^{-1/2}|\beta(0)|^{1/2} = (\tau_u\tau_v)^{-1/2}(f_v'g_u' - f_u'g_v')^{1/2}. \tag{40}$$

Using wavevector $k = 0$ in (39) yields

$$\tau_u g_v' + \tau_v f_u' > 0, \tag{41}$$

which can be finally written as

$$g_v' > -\eta f_u'. \tag{42}$$

Here $\eta = \tau_v/\tau_u$ is the ratio of characteristic response times. Recalling the expressions for f_u' and g_v' in terms of the original parameters, the condition for the Hopf bifurcationtakes the form

$$K > 1 + \eta. \tag{43}$$

Notice that the threshold value of K diminishes with the decrease of η.

The Turing instability ($\omega = 0$) arises when β decreases through zero, while α remains positive. Using (37), this condition is expressed as

$$\beta(k) = k^4 D_u D_v - k^2(D_u g_v' + D_v f_u') + f_u'g_v' - f_v'g_u' < 0. \tag{44}$$

There are two extrema points for the function $\beta(k)$: $k = 0$ and $k = k_c = (2D_u D_v)^{-1/2}(D_u g_v' + D_v f_u')^{1/2}$. When $D_u g_v' + D_v f_u' > 0$ (always possible with a suitable choice of parameters), k_c is well defined and corresponds to the minimum of β. The condition (44) for the Turing instability of a mode with wavevector k_c is given by

$$g_v' > -\varepsilon f_u' + 2\varepsilon^{1/2}(f_u'g_v' - f_v'g_u')^{1/2}, \tag{45}$$

or in terms of (35)

$$K > 1 + \varepsilon + 2\varepsilon^{1/2}\left[1 + \frac{3\pi^2}{4} - K\left(1 + \frac{\pi}{2}\right)\right]^{1/2}, \tag{46}$$

where $\varepsilon = D_v/D_u$ is the ratio of diffusion rates for the two subsystems.

Because $f'_u = -1 < 0$, conditions (42) and (45) can be satisfied only for $g'_v(\bar{u}, \bar{v}) > 0$, which is equivalent to the requirement that the uniform state should lie on the middle branch of the nullcline for the function g. In the optical model for the two-component system just described, it is easy to "tune" the parameters to obtain two possible types of instability and to alter the qualitative characteristics of the bifurcating solutions (such as ω_0 and k_c). Variations of η and ε, though less simple, are also possible. These possibilities are shown in Fig. 12, where the linear growth rate ν versus k is given for different external parameter values.

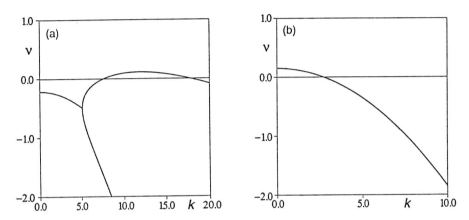

Fig. 12. Linear growth rate ν versus the modulus of the wavevector $\mathbf{k}$ for K greater than the Turing instability threshold (a) and the Hopf bifurcation threshold (b). (a) $K = 2.7$, $D_u = 0.02$, $D_v = 0.004$, $\eta = \tau_v/\tau_u = 3$. (b) $K = 2.3$, $D_u = D_v = 0.02$, $\eta = 1$.

The next question that naturally arises concerns behavior of the system above the first instability threshold. Near the onset of instability, i.e., when $|K^{th} - K|^{1/2} = \mu \ll 1$, it is possible to introduce the slowly varying amplitude $W(\mathbf{R}, T)$ ($\mathbf{R} = \mu\mathbf{r}, T = \mu^2 t$) for the solutions (34) as an order parameter. Use of perturbation methods leads to the Ginzburg-Landau equation for W [KuT]. Detailed discussion of these issues is beyond the scope of this article.

We conclude with several remarks concerning linear stability analysis. In the case of a Turing instability, linear stability considerations fix only the length of the critical wavevector, while its orientation remains arbitrary. This is a direct consequence of the rotational invariance of the problem. Due to nonlinearity, modes with different orientations of $\mathbf{k}$ become coupled. The product of two modes whose wavevectors have an angular separation of $\frac{2\pi}{3}$ is another mode with $|\mathbf{k}_3| = k_c$. Such spatial interactions can result in the formation of hexagonal patterns. Hexagonal patterns have been analyzed in a number of papers on nonlinear optical systems with a Kerr slice and feedback mirror, that is, for the case of rather strong diffraction [DAF]. It would be interesting to find out if the two-component model described above could also produce hexagons.

Consider an additional limitation to the understanding of nonlinear dynamics based on one result of linear stability analysis. Assume that the set of external parameters corresponds to the situation shown in Fig. 11b. Linear stability analysis results imply that the uniform state is stable with respect to small perturbations. However, the trivial state is not the only stable attractor for the system. Analysis shows that a multitude of various spatially localized patterns exist that can be either static or dynamic, such as traveling pulses [KeO]. The dynamics involve the "smallness" of either $\varepsilon = D_v/D_u$ or $\eta = \tau_v/\tau_u$ (or both) in a fundamental way. For insight into this problem, numerical simulation results for a one-dimensional case are provided for (28). Figure 13 is a snapshot of a traveling pulse obtained when $\eta = 0.0125$.

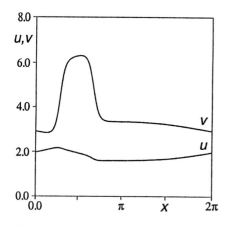

Fig. 13. A stationary traveling pulse. The parameters correspond to Fig. 11b.

4 Conclusion

We have examined various ways of designing complex spatio-temporal dynamics using nonlinear optical techniques. The overall architecture for spatial complexity was constructed by applying well-known models for multicomponent reaction-diffusion systems. We found that the optical design kit not only duplicated the significant concepts of classical synergetics, it had additional features as well. Now consider several sources of spatio-temporal complexity in optics.

1) Time delay in optical feedback. Involving time delay in the theory of differential equations, gives rise to new, interesting problems, and even trivial equations have nontrivial behavior, requiring the development of new mathematical methods. In a time-delayed system, the transition to chaotic dynamic regimes is a natural process; the real issue becomes finding a time-delayed system without chaotic dynamics. Time-delayed spatially distributed nonlinear systems essentially form the basis of a whole new field. What is the relationship between the dynamics of time and space? Only a few publications exist for systems with

spatial effects governed purely by temporal characteristics [BRT, IrV, Raz]. Due to the slow time response of the LCLV, it is possible to include elements, such as a computer-controlled TV camera with monitor, in the optical feedback loop of the optical synergetic block to control time-delayed feedback. Another simple method to include time delay is to place a time-delay element in the system's electronic feedback.

2) Long-range interactions in 2-D feedback. An optical system with 2-D feedback offers unique opportunities to design new types of spatial interactions by having a point in a laser beam cross section interact not only with its nearest neighbors, but with distant points as well [Vor]. Time delay is the simplest example of long-range interactions. However, "spatial delay," even for the simple case of field rotation in the optical feedback, changes the spatio-temporal dynamics radically. Using a complex spatial mapping in optical feedback is one of the more interesting techniques for developing more complex spatio-temporal dynamics. New dynamical phenomena can be obtained by supplementing classical multicomponent reaction-diffusion type systems with long-range interactions [DeV].

3) New types of local nonlinear interactions. The diffusion mechanism of local spatial coupling is most popular now, due perhaps to its simplicity. In optics, diffraction provides the most interesting type of local coupling. By accounting for diffraction, we enter into the field of multicomponent systems, because the two characteristics of an optical field (phase and intensity, or the real and imaginary components) are closely linked. It is easy to organize this type of local interaction in the feedback circuit of an optical synergetic block (perhaps the more difficult problem is to avoid diffraction). Diffraction in nonlinear systems with feedback produces many interesting examples of spatial self-organization in the optical field: hexagonal patterns, optical vortices, and akhseals [DAF, Bra, Vo1]. Is it possible to obtain different kinds of local spatial interactions in an optical system? The optical design kit has the ability to control local interactions using linear Fourier optics in the 2-D feedback of the optical synergetic block.

4) Parametric excitation of spatio-temporal instabilities. Parametric excitation gave traditional radio physics so much that it is natural to borrow this fruitful idea for nonlinear spatio-temporal dynamics. Note that parametric spatially distributed nonlinear systems are now of great interest [Clt]. There are several ways to obtain parametric excitation using the optical design kit. One way is to apply a periodically modulated controlling voltage to the LCLV. Periodically modulating input intensity or the polarization direction also produces parametric excitation.

Having discussed several sources of spatio-temporal complexity in nonlinear optics, we arrive at the main question: "Why are we interested in this complexity"? There are at least two possible answers. The first was formulated in very short form by Prof. H. Haken: "Complex systems are a challenge for the researcher." This challenge is enough for the scientist, but perhaps it is not enough for those who give money to science. Those who give money to science prefer practical applications for everyday life, so perhaps they will be satisfied by the following (longer) answer (asking for money requires that long answers be given).

It seems that the more complex the dynamic system, the greater the information capability of that system. For instance, the human brain is an excellent example of a complex system that has very good practical, everyday applications. One can argue that complicated systems, such as the atmosphere and the ocean, don't use their information capability for solving the types of problems handled by the brain. This is true, which provides just one reason why we should study the various aspects of complexity separately, instead of complexity in general. When we artificially grow a complex dynamic system, it must have two important properties: It must be governed by strong competition that depends on initial conditions; and it must have multiple components with a different time response for each separate component, so that both memory and history can exist. Perhaps when designing complexity in our everyday lives, it would be worthwhile to encourage competition and historical memory and to be more tolerant of nonsimilar subsystems.

References

[VoL]   Vorontsov, M.A., Larichev, A.V.: Adaptive Compensation of Phase Distortions in Nonlinear System with 2-D Feedback. Proc. SPIE **1409** 1991, 260-266.

[Vor]   Vorontsov, M.A.: Problems of Large Neurodynamics System Modelling: Optical Synergetics and Neural Networks. Proc. SPIE **1409** 1991, 116.

[Vas]   Vasiliev, A.A., Casasent, D., Kompanets, I.N., Parfionov, A.V.: Spatial Light Modulators. Radio i Sviaz, Moscow, 1987.

[Hak]   Haken, H.: Advanced Synergetics. Springer-Verlag, New-York, Berlin, 1987.

[Grs]   Grossberg, S.: Nonlinear Neural Networks: Principles, Mechanisms and Architectures. Neural Networks **1** 1988, 17.

[NiP]   Nicolis, G., Prigogine, I.: Self-Organization in Non-Equilibrium Systems, Wiley, New York, 1977.

[VKS]   Vorontsov, M.A., Koriabin, A.V., Shmalhauzen, V.I.: Controlling Optical Systems. Nauka, Moscow, 1988.

[Fis]   Fisher, R.A.: The Wave of Advance of Advantageous Genes. Ann. Eugenics **7** 1937, 355.

[KPP]   Kolmogorov, A.N., Petrovskii, I.S., Piskunov, N.S.: Vestn. Mosk., Univ. Mat. Mekh. **1** 1937, 1.

[AkV]   Akhmanov, S.A., Vorontsov, M.A., Ivanov, V.Yu., Larichev, A.V., Zheleznykh, N.I.: Controlling Transverse-Wave Interactions in Nonlinear Optics: Generation and Interaction of Spatiotemporal Structures. JOSA B **9** 1992, 78.

[VZI]   Vorontsov, M.A., Zheleznykh, N.I., Ivanov, V.Yu.: Opt. Quant. Electron **22** 1990, 501.

[AVI]   Akhmanov, S.A., Vorontsov, M.A., Ivanov, V.Yu.: Large-Scale Transverse Nonlinear Interactions in Laser Beams: New Types of Nonlinear Waves. JETPL **47** 1988, 707.

[FiW]   Fisher, A.D., Warde, C.: Opt. Lett. **8** 1983, 353.

[VKN]   Vorontsov, M.A., Katulin, V.A., Naumov, A.F.: Wavefront Control by an Optical-Feedback Interferometer. Opt. Comm. **71** 1989, 35.

[Kob]   Kobzev, E.F., Vorontsov, M.A.: Optical Implementation of "Winner Take All" Models of Neural Networks. Proc. SPIE **1409** 1991.

[And] Andronov, A.A., Leontovich, E.A., Gordon, I.I., Mayer, A.G.: Theory of Dynamic Systems on a Plane. Israel Program of Scientific Translations, Jerusalem, 1973.

[Tur] Turing, A.M.: Phil. Trans. Roy. Soc. B **237** 1952, 37.

[DAF] D'Alessandro, G., Firth, W.J.: Hexagonal Spatial Patterns for a Kerr Slice with a Feedback Mirror. Phys. Rev. A **46** 1992, 537.

[KeO] Kerner, B.S., Osipov, V.V. In: Selforganization by Nonlinear Irreversible Processes. Ed. Ebeling, W., and Ulbricht, H., Springer-Verlag, New-York, Berlin, Heidelberg. Springer Series in Synergetics, 1986.

[BRT] Le Berre, M., Ressayre, E., Tallet, A.: Self-Oscillations of the Mirrorlike Sodium Vapor Driven by Counterpropagating Light Beams. Phys. Rev. A **43** 1991.

[KuT] Kuramoto, Y., Tsuzuki, T.: On the Formation of Dissipative Structures in Reaction Diffusion Systems. Prog. Theor. Phys. 3 **54** 1975, 687.

[Krm] Kuramoto, Y.: Chemical Oscillations, Waves and Turbulence. Springer-Verlag, Berlin, Heidelberg, New York, Tokyo, 1984.

[Mar] Mardsen, J.E., McCracken, M.: The Hopf Bifurcation and Its Applications. Springer-Verlag, New York, Heidelberg, Berlin, 1976.

[Gbs] Gibbs, H.: Optical Bistability: Controlling Light with Light. Academic Press, Orlando, 1985.

[VRY] Vasiliev, V.A., Romanovski, Yu.M., Yakhno, V.G.: Autowave Processes. Nauka, Moscow, 1987.

[IrV] Iroshnikov, N.G., Vorontsov, M.A.: Transverse Rotating Waves in the Nonlinear Optical System with Spatial and Temporal Delay. In Essay in Nonlinear Optics: In Memoriam of Serge Akhmanov. Eds. Walter, H., Koroteev, N., and Scully, M.: IOP, London, 1992.

[Raz] Razgulin, A.V.: Rotational Multipetal Waves in Optical System with Feedback. SPIE Proc., 1993.

[DeV] Degtyarev, E.V., Vorontsov, M.A.: Spatial Dynamics of a Two-Component Optical System with Large-Scale Interactions. Nonlinear Optics **3** 1992, 295.

[Bra] Brambilla, M., Battipede, F., Lugiato, L.A., Penna, V., Prati, F., Weiss, C.O.: Phys. Rev. A **43** 1991, 5090.

[Vol] Vorontsov, M.A.: "Akhseals" as a New Class of Spatio-Temporal Light Field Instabilities. Sov. Quant. Electron. **20** 1993, 319.

[Clt] Coullet, P. in Proc. SUSSP, 1992.

3 Pattern Formation in Passive Nonlinear Optical Systems

W. J. Firth

Department of Physics and Applied Physics, University of Strathclyde, Glasgow, G4 ONG UK.

Introduction

Hexagons and other regular patterns are predicted in a number of wide-aspect nonlinear optical systems. Analysis and simulations of pattern formation in two classes of systems are presented. These are feedback systems, in which the mechanism is related to the Talbot effect; and cavity systems, where a tilted-wave mechanism operates. In the latter case, polarization instabilities are predicted to occur and show patterns. Gaussian beam pumping is the practical situation, and analysis and simulations with finite beam widths are also illustrated. From an applications viewpoint, defects in regular patterns are of potential interest. An "isolated state" memory is discussed as an example.

Information **is** pattern! All the information we possess, or at least all the information we have acquired, has been perceived as pattern. Arguably the most important channel is visual, i.e., optical patterns, essentially in a two-dimensional format. Thus two-dimensional optical patterns inevitably play a key role in our understanding and application of information transport, storage, and processing.

In one view – the reductionist approach – information is simply bits, and patterns are simply a collection of bits. This approach works well for simple cases, but seems to have limitations for complex operations, such as image processing. Our brains, on the other hand, seem to be very adept at such problems, apparently through a very effective process of pattern detection. Indeed we seem to actively seek pattern where none exists (e.g., the Man in the Moon), and science is in large measure a search for pattern and predictability in the world around us.

In this article, we describe some of the pattern formation and related processes recently observed in passive nonlinear optical systems. Passive means that the excitation is by an external driving field, smooth and constant in the ideal case, rather than through population inversion. Further, we will consider some of the defect structures and dynamic behavior also observed, arguing that these higher excitation phenomena may hold the key to the practical application of these effects in information processing. Indeed it could be argued that true (i.e., biological) neural systems actually operate in such a mode, rather than in the near-threshold regime, which characterizes most analyses of optical and other pattern-forming systems.

Much of the discussion will turn on analysis and simulations, because experimental confirmation of these phenomena is still somewhat limited. This is mainly due to the difficulty of achieving a large enough power over a large enough area to see fully developed patterns. Nevertheless, it is encouraging that most of the models discussed have some positive experimental evidence in their support and also that this body of evidence is growing at quite a rapid rate.

Most of the systems to be studied employ a third-order optical nonlinearity, of which the optical Kerr effect is the prototype. In a Kerr medium, the refractive index is intensity dependent, increasing (or decreasing) in proportion to the local optical energy density. The wave equation in a Kerr medium is thus nonlinear – of cubic order – in the field amplitude. In principle, all materials exhibit a Kerr effect, though it is generally too weak to be of practical use. Many materials exhibit a Kerr-like effect mediated by an electronic or molecular excitation: If the excitation is long-lived, a relatively weak light field can generate a large excitation, and thus induce a substantial index change. As a consequence, such systems exhibit a large effective nonlinearity, and these systems have been extensively investigated for a number of effects, including pattern formation. Among these materials are semiconductors, photorefractive media, and liquid crystals. While these media all have particular properties and features, their pattern-forming properties are broadly Kerr-like, and the details of their deviations from Kerr media need not concern us here.

An alternative and very effective approach is to synthesize a Kerr-like nonlinearity. This can be done by a transducer, which measures the optical intensity and induces a proportional refractive index change in an appropriate element of the system. This is strikingly successful in the case of liquid crystal light valves (LCLVs), where a very large Kerr effect can be synthesized locally over an area of several square centimeters, making it ideal for studies in pattern formation.[1]

We will examine systems in which the nonlinear medium sees itself in a single mirror (Sect. 2), and in which it is enclosed in a cavity (Sect. 3). The single mirror system is treated first, both because it has recently been used for a number of successful experiments and because it is functionally closely similar to the LCLV configuration just mentioned. The cavity configuration is attractive from an analytical and computational viewpoint and also offers some interesting phenomenology. It was already well developed in the field of optical bistability [Lug, Gib], and is well understood from an experimental viewpoint. It is worth noting that Moloney and coworkers found striking pattern formation and complex dynamics in simulations of passive cavity systems more than a decade ago [Gbs, MHG]. Other systems and configurations in this field are also of interest. For example, the simple situation of two optical beams counter-propagating in a Kerr medium has been found both experimentally and theoretically to give a very rich spectrum of patterns [Gry, BVS, FP, PPM, GIM]. More recently, it has been

---

[1] Ed. Note: Synthesis of Kerr-like nonlinearity using LCLVs and early experimental systems are more fully described in [VKS].

proposed that optical parametric oscillators, which are of great practical and fundamental interest, may display spatial instability and pattern formation [BL]. These latter devices use a second-order nonlinearity, which is present only in a restricted class of materials, and thus are qualitatively distinct from Kerr-like effects. Also different are active media, such as in lasers [TQG, LPN, AGR, JMN]. In lasers, there is no externally applied optical field to provide a phase reference, a difference that profoundly changes the allowable categories of pattern formation, because of the different symmetry properties of such systems.

A survey of work through 1989, describing transverse effects in lasers and nonlinear optics, has been published as foreword to a collection of papers in the field [AF]. As it happens, this just about marked the beginning of an upsurge of work on pattern formation in nonlinear optics. The articles by Lugiato [Lgi] and Weiss [Wei] in the Proceedings of the 1991 Solvay Conference review the early stages of that development. Further developments are described in a special issue of *Chaos, Solitons and Fractals* [Lia].

1 Induced and Spontaneous Patterns

How is an optical pattern created? A *kaleidoscope* is an example of a linear pattern generator. The apparent complexity of its output images is simply a projection of the arrangement of its parts and the external illumination thereof. The much richer alternative in optics as, for example, in biology, is to endow a simple system with an appropriate combination of free energy and nonlinearity, and allow it to organize itself into a complex structure. As we will see, the emergence of complex patterns and dynamics does not demand that either the properties or the driving of the system be complicated.

The concept of complexity emerging out of simple laws obeyed by simple systems is the main theme of this work. The key to this is nonlinearity. Nonlinear problems are not usually analytically solvable, so that either approximation or computational methods, or both, are required. Computer simulations figure strongly in this article, but are developed along with various analytic approaches. This is because computer simulations and analytic work complement each other. The analysis guides choices of problem and parameters for simulations, while the simulations, particularly in animated graphical form, can call attention to structures and behaviors that are not immediately apparent in the system equations.

The difficulty of nonlinear problems has compensating advantages. The operation of a linear system is wholly determined by its initial state and its environment, whereas the behavior of a nonlinear system depends upon its own state: It "looks at itself." This is literally true in the case of the nonlinear optical systems with feedback discussed in the following section. Nonlinear systems can seek out and maintain essentially the same optimum state in response to a wide variety of external conditions – a desirable device attribute.

Patterns can emerge *spontaneously* only in nonlinear systems; such patterns involve symmetry breaking. For example, a regular hexagonal lattice of bright spots has only discrete translational symmetry, so its emergence from a homogeneous state involves a breaking of that state's continuous translational symmetry.

Each symmetry breaking involves a choice among two or more equivalent broken-symmetry states: The set of such states retains the underlying symmetry, but the system chooses only one. Coexistence of two (or many) states consistent with the same input environment is the key distinction between linear and nonlinear optical patterns. In particular, it offers a possible route to applications, because coexistent states can form the basis of all-optical memory devices.

Note that one need not restrict oneself to stable states in identifying states with bits of information. Dynamic, chaotic, or even unstable states can be used provided they can be accessed, can be differentiated from competing coexistent states, and are robust against noise or other perturbations. More subtly, it is often possible to stabilize unstable states with small feedback control. Such techniques are often termed control of chaos [Roy]. Controlled unstable states have a number of advantages, some of which have already been demonstrated in optics. These techniques are outside the scope of this article, but I mention them to emphasize that unstable states are not irrelevant and because finding techniques to select and stabilize unstable spatial states (e.g., images) is an important and exciting challenge in this field.

1.1 Materials and Geometries

Nonlinear optics of the "traditional" kind – primarily frequency conversion – is passive, in the sense that the medium itself is not excited, nor does it have a dynamic response.[2] In fact, it plays only an enabling role for the transfer of energy between optical fields. One can consider a more general form of nonlinear response in which optical energy is absorbed by the medium and affects its optical properties, so that the fields and the medium play a much more equal role. Indeed the phenomena are often dominated by the dynamical response of the medium's excitation, especially in cases of resonance. Media, such as liquid crystals [AVI, KBT, ME, TBW], photorefractives [AGR, Hon, Ban], alkali vapors [GVP, BVS, PPM], and semiconductors [GMG, MSJ], have relatively slow responses. For such media, relaxation and transport processes become important, and a coupled-matter field approach becomes necessary.

In the context of optical pattern formation and stability, it is usually the slowest response time of the system that dominates. For liquid crystals and photorefractives, this can be slow enough to observe the dynamics directly in real time, directly or with a video recorder. While these systems have limitations on applicability, this makes such systems ideal for fundamental investigations and exploration.

Turning from the medium to the geometry, any regular or quasi-regular pattern has a characteristic length scale. For an ideal spontaneous pattern, this scale is determined neither by the system size nor the nonlinear material, but emerges from the dynamic process itself. We will examine two distinct ways in which the characteristic size may be determined.

---

[2] Ed. Note: Except perhaps in the case of thermal blooming in laser beams, an effect discovered at the advent of nonlinear optics [LMW, AKM].

In the following section, we examine feedback systems in which a diffraction length determines the scale. A Kerr-like medium effects only the phase of the optical field. To close the feedback loop, it is necessary to convert that phase modulation to amplitude modulation. Diffraction does just that, at distances of one-quarter and/or three-quarters of the Talbot length. The latter is the length at which a periodic pattern self-images as it propagates at points where the different spatial Fourier components have relative phase shifts, which are multiples of 2π.

In the subsequent section, we examine pattern formation in optical cavities. If the cavity round-trip is not an exact number of optical wavelengths, then excitation of on-axis modes will be suppressed. If the cavity mistuning has the appropriate sign, however, an off-axis or tilted wave can be cavity resonant. The transverse wavevector necessary to cancel the mistuning then determines the scales of the pattern. This mechanism is widely applicable, operating in lasers ([JMN] and [BL]), as well as the passive cavity systems discussed below.

For applications, high spatial resolution is normally desirable. Informatic processes based on spontaneous patterns are obviously scaled in proportion to the pattern, which should thus be as fine as possible. The particular models discussed are limited by the paraxial approximation to transverse scales of many wavelengths, i.e., at least $10\mu m$ or so. The two mechanisms need not be so limited, however, and it may well prove possible to extend these configurations beyond the paraxial limit, maybe even to the $1\mu m$ scale. This is still quite large by microelectronic standards, but the problem for optical systems is not just packing density, but also power density, and the latter improves by two orders of magnitude for every order of magnitude reduction in pattern scale. Of course, the medium may provide a limitation, if through diffusion or similar processes, it presents a finite spatial bandwidth, but this is an engineering problem rather than a fundamental limitation.

2 Mirror Feedback Systems

Formation of transverse structures requires nonlocal interactions in the nonlinear optical medium. In the cavity systems described below, this is accomplished by the action of diffraction on the circulating cavity field. Through interference at the input mirror, both amplitude and phase of the cavity field are modulated by the input field. In this section, feedback affects the phase only, and no energy is recirculated.

The first case to be examined is that of a thin slice of Kerr medium, illuminated from one side by a plane wave (or smooth) input field. The transmitted field propagates to a mirror. The reflected portion returns and re-enters the Kerr slice and modulates the phase of the input field *cross-phase modulation*. As we will see, this cross-modulation is the essential element in pattern formation. Because the feedback field has no other effect on the input field, it is immaterial whether the cross-modulation is applied through a true or a simulated Kerr effect. The latter can, as already noted, be very simply and effectively implemented with a liquid crystal light valve (LCLV). The LCLV has a much wider range of behaviors and potential than can be described here, and the interested reader should consult other articles for more information and references.

2.1 Kerr Slice with Feedback Mirror

An anti-reflected slice of Kerr medium coupled to a feedback mirror is a very simple optical system, but nevertheless presents very complex spatio-temporal properties. Here, as in most such problems, one begins with a linear stability analysis of the unpatterned state. This analysis can be carried forward most simply and completely for the case in which the input field is a plane wave, i.e., with spatially homogeneous intensity and flat phase fronts, aligned parallel to the Kerr slice, assuming normal incidence. Considering transverse degrees of freedom arising from diffraction in the propagation to and from the mirror, and additionally allowing for diffusion of the excitation within the slice, Firth [Frh] presented a linear stability analysis of this system, predicting both static and dynamic instabilities. The basic mechanism arises from the fact that a beam of uniform amplitude, but with a non-uniform phase will develop an amplitude variation as it propagates, due to diffraction effects. As mentioned above, this is related to the Talbot effect.

Suppose that the feedback field at the Kerr slice has an amplitude variation, such as a hexagonal intensity pattern. It will impose a corresponding phase modulation on the input field in the slice. Propagation to the mirror and back will, in general, at least partially convert this phase modulation to amplitude modulation. When this amplitude variation reproduces the initiating one, then we have the possibility of a self-sustaining stable pattern. These features of optical propagation have been known since the early days of the wave theory of light. In 1836, Talbot showed that a light field with a regular pattern tends to self-image after a characteristic distance now known as the Talbot length L_T. More importantly for our present purposes, a pattern that has amplitude variations, but uniform phase (AM), or vice versa (PM), cycles through states of pure AM and pure PM as it propagates. The conversion distance is just one-quarter of the Talbot length, so that, in particular, a pure PM beam, such as a plane wave having traversed a thin Kerr slice, will become pure AM after any odd multiple of $L_T/4$. If that light is fed back into the Kerr slice, e.g., by a feedback mirror, this AM feedback beam will phase modulate the input field and thus close a feedback loop, which can lead to pattern formation.

More generally, the pattern might replicate only after two or more round-trips, which leads to oscillatory patterns. If the Kerr effect arises through a material excitation with a finite lifetime, then the slice has a memory, and so need not oscillate at periods integrally related to the round-trip time. More significantly, if the lifetime is long compared to the round trip-time, then only static patterns, or dynamic patterns on time scales comparable to the excitation lifetime, are possible.

These basic features are apparent from the linear stability analysis [Frh]. The nature and properties of the patterns that arise can only be determined by nonlinear analysis or simulation. Nonlinear analysis and numerical simulations in two transverse dimensions were subsequently reported [AF, AlF] showing, in particular, the spontaneous appearance of hexagonal patterns. Employing a liquid crystal cell as Kerr slice, Macdonald and Eichler [ME] demonstrated pattern formation with spatial scales in accord with the theory, but without clear evidence of hexagonal structure. Tamburrini and colleagues [TBW] did report hexagons in a broadly similar system, while Grynberg and colleagues [GMP] observed hexagonal and other structures in a rubidium vapor cell with feedback mirror. Honda [Hon] and Banarjee and colleagues [Ban] report hexagons in photorefractive crystals with reflective feedback, but with a more complex interplay of nonlinearity and diffraction.

In this abbreviated treatment, we give some typical numerical results for this problem and describe a perturbation approach to the nonlinear behavior of the system. In particular, we present amplitude equations of Ginzburg-Landau type for the problem, which are of a form typical in hydrodynamics [CCL, Pom].

2.2 Basic Model and Stability Analysis

Consider a thin anti-reflected slice of Kerr medium irradiated from one side by a spatially smooth beam, indeed a plane wave for the moment. A plane feedback mirror behind generates a counter-propagating beam in the Kerr slice (see Fig. 1) and closes the feedback loop. No energy is stored in the system, the complexity arises from the cross-phase modulation of the two beams in the Kerr slice.

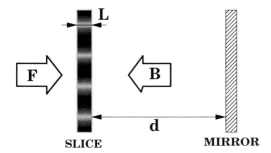

Fig. 1. Schematic diagram of the model. A thin slice of Kerr material of thickness L is illuminated from the left. The mirror, at a distance d, reflects this field back through the slice, thus closing the feedback loop.

The system that we have modeled is a good approximation of a medium whose refractive index depends on the density of some optically driven excitation. In the simplest case, the equation for the excitation density reads

$$-l_D^2 \nabla_\perp^2 n + \tau \frac{\partial n}{\partial t} + n = |E|^2.$$

The total electric field is the sum of a forward and backward field

$$E = F(\bar{\mathbf{x}}, t)e^{i(k_0\bar{z}-\omega_0 t)} + B(\bar{\mathbf{x}}, t)e^{i(k_0\bar{z}+\omega_0 t)} + \text{c.c.}.$$

where F and B are slowly varying amplitudes. We assume that the excitation density changes (linearly) the refractive index of the medium, so that the electric field is phase modulated during the transit of the slice. We neglect absorption for simplicity. We suppose that the slice is sufficiently thin, so as to allow the neglect of diffraction inside it. Under all these assumptions, Maxwell's equations for the two fields have a very simple form within the slice [AlF]

$$\frac{\partial F}{\partial \bar{z}} = i\chi_0 n \qquad \frac{\partial B}{\partial \bar{z}} = -i\chi_0 n B, \tag{1}$$

while the equation for the excitation density is

$$-l_D^2 \nabla_\perp^2 n + \tau \frac{\partial n}{\partial t} + n = |F|^2 + |B|^2. \tag{2}$$

The modulus of χ_0 measures the strength of the nonlinearity, while its sign determines the type – positive for focusing, negative for defocusing media.

Finally, outside the material the fields' evolution is determined by Maxwell's equations in free space; for example, the forward field evolves according to

$$\frac{\partial F}{\partial \bar{z}} = \frac{i}{2k_0} \nabla_\perp^2 F. \tag{3}$$

As in Fig. 1, the mirror distance is d.

These equations admit a spatially uniform equilibrium

$$|F|^2 = I_0, \quad |B|^2 = RI_0, \quad n = I_0(1 + R), \tag{4}$$

where I_0 is the intensity of the forward (or pump) field and R is the reflectivity of the mirror. Define $\sigma \equiv d/(k_0 l_D^2)$ as an adimensional parameter, which measures the relative strength of diffraction versus diffusion (σ is small for strong diffusion). Then this uniform solution is linearly stable with respect to a perturbation of wavevector $\mathbf{K}$ for forward field intensities that are smaller than the value of the pump, I_0 which satisfies

$$1 + K^2 - i\Omega = 2RI_0\chi \sin(\sigma K^2)e^{i\Omega t_R}, \tag{5}$$

where t_R is the round-trip time from slice to mirror and back, $K = |\mathbf{K}|$, and Ω is the oscillation frequency of the perturbation.

Equating the real and imaginary parts, we find that the oscillation frequency of the perturbation Ω is either zero or of the order of the round-trip time. Here we consider only slow media, so that dynamic instabilities ($\Omega \neq 0$) have a threshold much higher than the static instabilities ($\Omega = 0$) and can be ignored. Putting

$\Omega = 0$ in (5), we obtain the instability threshold for the uniform solution with respect to such a nonoscillating perturbation of wavevector $\mathbf{K}$

$$|\chi| I_0 = \frac{1 + K^2}{2R|\sin(\vartheta)|} \qquad \chi \sin(\vartheta) > 0. \tag{6}$$

For fixed σ, the value of ϑ for which the threshold is minimum, ϑ_{th}, is given by

$$\tan(\vartheta_{\mathrm{th}}) = \vartheta_{\mathrm{th}} + \sigma. \tag{7}$$

We can see that for small diffusion ($\sigma \gg 1$) the minimum instability threshold is at $\vartheta_{\mathrm{th}} \simeq \pi/2$ for a focusing medium $\chi > 0$, or at $\vartheta_{\mathrm{th}} \simeq 3\pi/2$ for a defocusing medium $\chi < 0$. For smaller σ, the threshold increases while the minima tend to smaller values of ϑ (see Fig. 2). These predictions have been confirmed in experiments using a liquid crystal as a Kerr-like medium [ME, TBW]. The latter authors were able to verify the formula for both signs of χ by taking advantage of the fact that a negative slice-mirror distance is equivalent to a sign change in χ. A negative effective mirror distance is achieved by using a lens combination to create a "virtual mirror," which can be on either side of the slice.

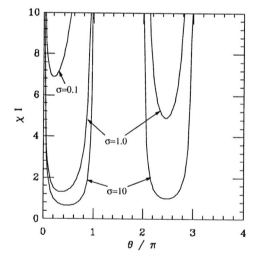

Fig. 2. Threshold curves for a focusing medium for three different values of σ. $R = 0.9$, $\chi > 0$.

Nonlinear Analysis. The linear analysis gives a threshold, which depends only on the magnitude of K and not on its direction. To predict which vector, $\mathbf{K}$ or vectors $\mathbf{K}_1$, $\mathbf{K}_2, \ldots$ will actually form the eventual pattern, one must undertake some form of *nonlinear analysis*. This is usually carried out by using singular perturbation theory to derive *amplitude equations* for the amplitudes of one or more *rolls*, i.e., $\mathbf{K}$s. In the present case, it is sufficient to analyze a set of three rolls at 60°. We can write the excitation density as

$$n = (Ae^{i\mathbf{K}_1\cdot\mathbf{x}} + Be^{i\mathbf{K}_2\cdot\mathbf{x}} + Ce^{i\mathbf{K}_3\cdot\mathbf{x}} + \text{c.c.})/2$$

and obtain three Ginzburg-Landau-type equations for the complex amplitudes

$$\frac{\partial A}{\partial t} = \mu A + \eta B^* C^* - [\zeta_1|A|^2 + \zeta_2(|B|^2 + |C|^2)]A$$

$$\frac{\partial B}{\partial t} = \mu B + \eta A^* C^* - [\zeta_1|B|^2 + \zeta_2(|A|^2 + |C|^2)]B \qquad (8)$$

$$\frac{\partial C}{\partial t} = \mu C + \eta A^* B^* - [\zeta_1|C|^2 + \zeta_2(|A|^2 + |B|^2)]C$$

$$\begin{cases} \mu = 2R\chi I_{\text{th}}\sin(\vartheta_{\text{th}})(p-1) \\ \eta = RI_{\text{th}}p\chi^2(1 - \cos(\vartheta_{\text{th}})) \\ \zeta_1 = \frac{1}{4}RI_{\text{th}}p\chi^3(3\sin(\vartheta_{\text{th}}) - \sin(3\vartheta_{\text{th}})) \\ \zeta_2 = \frac{1}{2}RI_{\text{th}}p\chi^3(2\sin(\vartheta_{\text{th}}) - \sin(2\vartheta_{\text{th}})). \end{cases}$$

Similar equations have been obtained in a number of fields of physics, from hydrodynamics [CCL, SS] to flame physics [ShS]. This should not be surprising as these equations describe the simplest form of interaction of three waves whose wavevectors satisfy the condition: $\mathbf{K}_1 + \mathbf{K}_2 + \mathbf{K}_3 = 0$.

The first term on the right of each member of (8) is the linear decay or growth of the wave. The second is a source term due to the resonance with the sum of the other two waves ($\mathbf{K}_1 = -\mathbf{K}_2 - \mathbf{K}_3$). This term only occurs for this triad resonance, and its presence thus suggests that hexagons will dominate close to threshold. All three waves need to be present for this term to be active, so it requires at least two transverse dimensions. Finally, the last term is the only form of third-order nonlinearity that is resonant. In the present problem both ζs are intrinsically positive, and the last term thus represents a saturation.

Note that the quadratic term is sensitive to the phases of the three amplitudes, while the third-order term is not. The quadratic term is thus vulnerable to noise or other influences that might perturb these phases. Additionally, any other wavevector present will contribute to the third-order term, but only one particular pair of directions contributes to the quadratic term. As a consequence, the emergence of the pattern from noise and the selection of the particular direction(s) of $\mathbf{K}$ that will survive are actually dominated by the third-order term, even if one would expect the lowest order nonlinear term (i.e., the quadratic) to dominate when all amplitudes are small [VFi].

It is possible to analyze the stationary states of this model exactly in the same way as in [CCL] (see Fig. 3). In this problem, the form of the coefficients poses some constraints on the solution types and their stability.

There are three kinds of stationary solutions:

(1) The uniform solution ($A = B = C = 0$) is stable for $p < 1$, unstable otherwise (line U in Fig. 3).

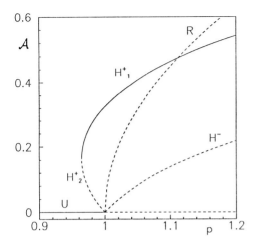

Fig. 3. Bifurcation diagram of (8). The continuous line indicates stable solutions, the dashed lines unstable solutions. $R = 0.9$, $\sigma = 10$, and $\chi = 1.0$.

(2) Rolls (curve R in Fig. 3) are given by

$$A = \mathcal{A}e^{i\varphi} \quad B = C = 0 \quad \begin{cases} \mathcal{A} = \dfrac{1}{|\chi|}\sqrt{\dfrac{8(p-1)\sin(\vartheta_{\text{th}})}{3\sin(\vartheta_{\text{th}}) - \sin(3\vartheta_{\text{th}})}} \\ \varphi \in [0, 2\pi) \end{cases}$$

and any cyclic permutations. They exist for $p > 1$, but are *always unstable*.

(3) Hexagons, given by

$$A = B = C = \pm\mathcal{A} \quad \mathcal{A} \in \mathcal{R}^+ \qquad p = \frac{1}{1 \pm T_1\mathcal{A} + T_2\mathcal{A}^2} \qquad (9)$$

$$\begin{cases} T_1 = \chi\dfrac{\sin^2(\vartheta_{\text{th}}/2)}{\sin(\vartheta_{\text{th}})} \\ T_2 = -\chi^2\dfrac{11\sin(\vartheta_{\text{th}}) + 4\sin(2\vartheta_{\text{th}}) + \sin(3\vartheta_{\text{th}})}{8\sin(\vartheta_{\text{th}})}. \end{cases}$$

The $\pm$ sign in the equation for p corresponds to the $\pm$ sign in the amplitudes. As T_1 and T_2 are, respectively, positive and negative independently of the sign of the nonlinearity, the bifurcation diagram (Fig. 3) is the same for both focusing and defocusing media. The hexagon solution exists only if

$$p > \frac{4T_2}{4T_2 + T_1^2}.$$

The $+$ sign in (9) gives the two curves H_1^+ and H_2^+ in Fig. 3. The upper branch is always stable, the lower branch is unstable. The $-$ sign, instead, gives the curve H^-; this solution is phase unstable. In this model, the H_1^+ curve is stable whatever the value of the pump.

Numerical study is necessary to get a more complete understanding and to check that the approximations done in the analytical study are correct. Because this model system is relatively simple, simulations can be run at a reasonable speed on desktop workstations. Details have been published elsewhere [AlF].

In the simulations, the initial condition was a random perturbation of the uniform solution. Near threshold this random noise grows until bright spots appear and arrange themselves on a hexagonal grid (Fig. 4). This behavior is typical for both a focusing and a defocusing medium. For increasing values of the pump, the hexagonal structure becomes unstable and the bright spots wobble around their position. If the pump is increased further, the motion becomes more pronounced until the system reaches a state of spatio-temporal chaos.

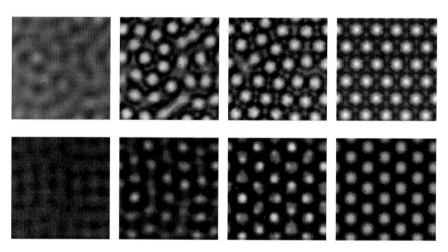

Fig. 4. Backward field intensity in a focusing (top) and a defocusing (bottom) medium. Gray scale, from white (high intensity) to black (low intensity). Time grows from left to right. $R = 0.9$, $t_R = 0.05$ in both cases. For the focusing medium, $\sigma = 1.0$ and $\chi = +1.0$; and $\sigma = 0.5$ and $\chi = -1.0$ for the defocusing case.

Hexagons and other patterns have been observed in Kerr-slice experiments [ME, TBW]. In experiments, the finite size of the beam can affect the pattern formation in a significant way. In particular, it means that the continuous translational symmetry of the plane-wave problem is reduced to a cylindrical symmetry about the optic axis. It is, therefore, necessary to extend the above model to deal with Gaussian beams. This has been done by a combination of symmetry arguments and numerical simulations [PaA]. For example, pentagonal patterns are both predicted and observed in certain parameter ranges. Fivefold-symmetric patterns cannot, of course, form infinitely periodic lattices in a two-dimensional space. Figure 5 shows such a structure [PaA]. Recent experiments have demonstrated most of the predicted features in both "intrinsic" [TBW] and LCLV [PRR] configurations. The latter are discussed below.

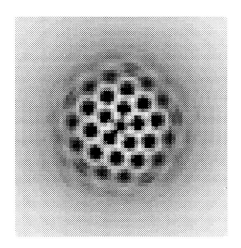

Fig. 5. Pentagonal structure found in the Kerr slice with feedback mirror [PaA] for the case of Gaussian-beam excitation.

In summary, this simple model clearly demonstrates that space-time complexity in nonlinear optics does not require complex material response. Basic and universal phenomena, such as the effect of propagation in converting phase modulation to amplitude modulation in conjunction with some nonlinearity, are already sufficient to produce a huge variety of beautiful and interesting phenomena.

2.3 Liquid Crystal Light Valve Systems

An LCLV with feedback forms a simple and practical system capable of displaying a dazzling range of spontaneous static and dynamic patterns [AVl, Vor]. This system includes within its repertoire of pattern formation a direct analogue of the Kerr slice with feedback mirror discussed above [FV][3].

Using the LCLV as an artificial Kerr slice, different nonlinear systems with two-dimensional feedback have been implemented [AVl, VMi, TNT, PRR]. In some of them, there is transformation of the phase modulation into intensity modulation by diffraction of the optical beam in the feedback loop. The spatio-temporal phenomena in these systems can be described by the following equation [Vor]:

$$\tau \frac{\partial u}{\partial t} + u = l_D^2 \nabla_\perp^2 u + K \cdot I_u(\mathbf{r}, t), \tag{10}$$

where $u(\mathbf{r}, t) = lk\Delta n(\mathbf{r}, t)$, $(k = 2\pi/\lambda,)$ l is the thickness of the liquid crystal layer and Δn is the change of refractive index caused by the controlling light intensity I_u on the photoconductor layer.

---

[3] Ed. Note: The principle of LCLV operation is described in Chapter 2.

After diffractive transformation, the intensity distribution on the photoconductor is given by

$$I_u = \mid F(\mathbf{r}, L, t) \mid^2, \tag{11}$$

where $F(\mathbf{r}, L, t)$ is the complex amplitude of the laser beam after diffracting over a distance L in the feedback loop. The complex amplitude at the beginning of the feedback is $F(\mathbf{r}, 0, t) = F_0 \cdot \exp iu(\mathbf{r}, t)$. The propagation from 0 to L is described by the parabolic equation (3) for the complex amplitude.

It is clear from comparison of (10, 2) and from (11) and (3) that the mathematical model of this feedback system is equivalent to the Kerr-slice-feedback-mirror model described in the previous section. The only significant difference, is that here the modulation is due only to the feedback field, whereas in the Kerr slice there is also a self-phase modulation by the input field. This is of no consequence for a plane-wave input, but for a Gaussian beam, it is approximately equivalent to a lens, positive or negative with the sign of the nonlinearity. This affects the propagation of the beam as a whole, such as in the case of a negative lens, where it can severely reduce the overlap of the reflected field with the input, and thus the potential for instability.

Hexagons and other patterns have been observed in LCLV feedback systems [Vor, TNT, PRA]. As in the "intrinsic" Kerr-slice experiments [ALV], the finite size of the beam can affect the pattern formation in a significant way. Recent experiments have demonstrated predicted features in LCLV configurations [PRA, ALV].

3 Pattern Formation in Optical Cavities

It has been known for some time that a nonlinear cavity driven by a coherent input field can show a number of nonequilibrium phase transitions, including optical bistability [Gib] and transverse pattern formation [MMN, LL, LO, QAM, MAI, HVR, Vit, FSM, HTW].

In particular, a mean-field model of a nonlinear cavity [FSM] was found to give rise to hexagonal patterns, at least for a self-focusing medium. Here we describe and extend that model, including polarization effects in the third-order nonlinear susceptibility $\chi^{(3)}$. It was found [GMW] that there are, in general, two pattern-forming modes, one of which preserves the linear polarization of the input field (symmetric mode), while the other does not (antisymmetric mode). The symmetric mode, which dominates in a self-focusing medium, gives rise to hexagonal patterns [FSM], and is essentially equivalent to the scalar field model (at least close to threshold). The antisymmetric mode may be dominant in a self-defocusing medium. In this case, rolls dominate close to the instability threshold, while further from equilibrium, a variety of irregular structures occur, some metastable and some pinned by the effects of the boundaries.

An important feature of these models concerns the mechanism underlying the pattern formation. Here, pattern formation occurs when the cavity is mistuned in such a direction that the wavevector of the light is larger than that of the nearby cavity mode. In such a case, off-axis or "tilted" waves can exactly match the cavity, and it is precisely this fitting requirement that determines the transverse wavevector of the most unstable mode. This is a geometrical and *linear* mechanism for determination of the scale of optical patterns, though naturally nonlinearity is required to endow the resonant tilted waves with gain. Here that gain arises through four-wave mixing.

Another attraction of these models is that they link with other models of optical pattern formation in a cavity. Recently, pattern formation has been predicted in optical parametric oscillators [BL]. In broad-area lasers, exact transverse traveling-wave solutions have been found [JMN] in a mean-field model. In both these cases, the transverse scale is determined by precisely the tilted wave mechanism we describe.

As with the Kerr-slice models described in the previous section, it turns out that the self-focusing (symmetric, scalar) Kerr cavity gives hexagons and only hexagons at third order in perturbation theory. For this reason, as well as for physical reasons, it is of interest to generalize to a two-level type of nonlinearity [FS]. This is Kerr-like in the limit of large atomic detuning, but becomes saturable and develops an absorptive component closer to atomic resonance. Pattern formation in this system is rather more varied, showing roll-hexagon competition [FS]. This system's properties will be outlined below. One potentially important property is the predicted existence of *isolated states*, which include single bright spots on a lower-amplitude flat background. Such states coexist with the uniform solution, and thus could be the basis of pixels for information processing.

3.1 Vector Kerr Model and Equations

Our basic model consists of a cavity, Fabry-Pérot or ring, which is filled with an isotropic Kerr medium and driven by a linearly polarized input field (see Fig. 6). This situation may be described by an appropriate generalization of the mean field model [LL, LO, FSM] to allow for the vector nature of the field. For an isotropic medium, the third-order nonlinear polarization can be written in the general form

$$\mathbf{P}^{(3)} = 3\varepsilon_0 \chi_{1111}^{(3)} \left(A(\mathbf{E} \cdot \mathbf{E}^*)\mathbf{E} + \frac{B}{2}(\mathbf{E} \cdot \mathbf{E})\mathbf{E}^* \right), \tag{12}$$

where $\mathbf{E}$ is the positive frequency component of the vector electric field, and

$$A = (\chi_{1122}^{(3)} + \chi_{1212}^{(3)})/\chi_{1111}^{(3)} , \ B = 2\chi_{1221}^{(3)}/\chi_{1111}^{(3)}.$$

For an isotropic medium, $A + B/2 = 1$. For example, for the Kerr effect in liquids $A = 1/4$, $B = 3/2$. Atomic vapors, while not Kerr media, are likely to show similar phenomena. They offer considerable flexibility in both the sign of $\chi^{(3)}$ and the relative magnitudes of A and B. For example, the $D1$ transition in sodium vapor ($J = 1/2$ to $J = 1/2$) yields $A = 1/2 + \beta/4$ and $B = 1 - \beta/2$ in the Kerr limit, where β is the optical pumping parameter [MB].

In previous work, linearly polarized fields were considered, in which case (12) reduces to the scalar relation $P^{(3)} = 3\varepsilon_0 \chi_{1111}^{(3)} |E|^2 E$, leading to the nonlinear term in the mean-field-evolution equation in [LL].

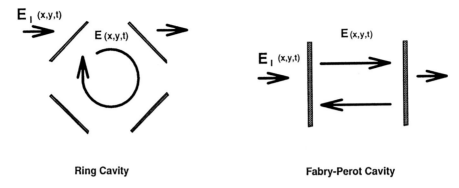

Ring Cavity   **Fabry-Perot Cavity**

Fig. 6. The configuration of the nonlinear cavity, either in ring or Fabry-Pérot geometry. $\mathbf{E}_I$ is the input field and $\mathbf{E}$ is the internal cavity field.

Based on (12), we can generalize the mean-field model to allow for field polarization, and the evolution equation for the electric field becomes

$$\frac{\partial \mathbf{E}}{\partial t} = -(1 + i\eta\theta)\mathbf{E} + \mathbf{E_I} + ia\nabla^2\mathbf{E} + i\eta\left(A(\mathbf{E} \cdot \mathbf{E}^*)\mathbf{E} + \frac{B}{2}(\mathbf{E} \cdot \mathbf{E})\mathbf{E}^*\right), \quad (13)$$

where $\mathbf{E} = \mathcal{E}_x\hat{\mathbf{x}} + \mathcal{E}_y\hat{\mathbf{y}}$ is the (scaled) vector electric-field envelope, $\mathbf{E}_I$ is the input field, $\eta = +1(-1)$ indicates self-focusing (self-defocusing), θ is the cavity detuning parameter, ∇^2 is the transverse Laplacian, and a measures the relative strength of transverse diffraction. The scaling employed is identical to that used in references [LL, LO, FSM]. The scalar field model is recovered in the limit of a linearly polarized field, e.g., $\mathbf{E} = \mathcal{E}\hat{\mathbf{x}}$ for arbitrary A and B.

To proceed, we transform the vector equation (13) to a circularly polarized basis defined by

$$\mathcal{E}_\pm = \frac{1}{\sqrt{2}}\left(\mathcal{E}_x \pm i\mathcal{E}_y\right)$$

to obtain the coupled evolution equation

$$\frac{\partial \mathcal{E}_\pm}{\partial t} = -(1 + i\eta\theta)\mathcal{E}_\pm + ia\nabla^2\mathcal{E}_\pm + \mathcal{E}_o + i\eta\left(A|\mathcal{E}_\pm|^2 + (A+B)|\mathcal{E}_\mp|^2\right)\mathcal{E}_\pm, \quad (14)$$

which is specialized to the case in which the input field is linearly polarized along the x-direction, i.e., $\mathbf{E}_I = \sqrt{2}\mathcal{E}_o\hat{\mathbf{x}}$. Note that in the case of pure circularly polarized input (14) is identical to the scalar model up to a rescaling of the field amplitude. Indeed in the latter case, the opposite circular polarization can never show an instability, so a scalar description suffices.

3.2 Spatial Stability of Symmetric Solutions

The symmetric steady-state plane-wave solution of equation (14) obeys

$$\mathcal{E}_o = \mathcal{E}_s\left(1 - i\eta(2I_s - \theta)\right). \quad (15)$$

To examine the stability of this solution, we set

$$\mathcal{E}_{\pm}(x, y, t) = \mathcal{E}_s \left(1 + \psi_{\pm}(x, y, t)\right), \tag{16}$$

which, on substituting into equation (14), yields

$$\frac{\partial \psi_{\pm}}{\partial t} = -\left(1 + i\eta(\theta - \eta a \nabla^2 - 2I_s)\right) \psi_{\pm} \tag{17}$$
$$+ i\eta I_s \left(A(\psi_{\pm} + \psi_{\pm}^* + |\psi_{\pm}|^2) + (A + B)(\psi_{\mp} + \psi_{\mp}^* + |\psi_{\mp}|^2)\right)(1 + \psi_{\pm}).$$

We seek solutions of the form

$$\mathbf{U}(x, y, t) = \mathbf{u} \exp(i\mathbf{K} \cdot \mathbf{x} + \sigma t), \tag{18}$$

where $\mathbf{K} = (K_x, K_y)$ and $\mathbf{x} = (x, y)$. For such solutions to exist, σ must satisfy one of the following equations:

$$(\sigma_1 + 1)^2 + (\theta_K - 6I_s)(\theta_K - 2I_s) = 0, \tag{19}$$

$$(\sigma_2 + 1)^2 + (\theta_K + 2(B - 1)I_s)(\theta_K - 2I_s) = 0, \tag{20}$$

where $\theta_K = \theta + \eta a K^2$, and σ_1 and σ_2 are eigenvalues of the linear stability matrix [GMW]. One eigenvalue corresponds to a symmetric mode (S-mode), while the other corresponds to an antisymmetric mode (A-mode).

If the S-mode prevails, then the field will remain linearly polarized even though the plane-wave symmetric solution is unstable. This case is identical to the scalar case [LL, FSM]. The neutral stability curve, where the S-mode is marginally stable ($\sigma_1 = 0$), is given by

$$I_s = \frac{1}{3} \left(\theta_K \pm \frac{1}{2}\sqrt{\theta_K^2 - 3}\right). \tag{21}$$

This equation demonstrates the "tilted-wave" mechanism, in that it is clear that the threshold depends only on θ_K, i.e., the "tilt" associated with finite K adds to the cavity mistuning θ, and can annul it if θ is *negative*.

For a self-focusing medium, $\theta_K > \sqrt{3}$ can always be achieved for some K. For $\eta = -1$, however, $\theta_K < \theta$ and (21) can be satisfied only for $\theta \geq \sqrt{3}$, i.e., in the bistable region with I_s already plane-wave unstable.

If the A-mode dominates, then the instability causes the vector field to evolve away from the linear polarization state of the input field. In general, it becomes elliptically polarized. The neutral stability curve, where the A-mode is marginally stable ($\sigma_2 = 0$), is given by

$$I_s = \frac{\theta_K(B - 2) \pm \sqrt{\theta_K^2 B^2 + 4(B - 1)}}{4(B - 1)}. \tag{22}$$

Though the system may become unstable to both the S-mode and the A-mode, the neutral stability formula of equations (21) and (22) show that the S-mode is always the first to become unstable in a self-focusing medium. The instability is to a finite wave number if $\theta < 2$. An example of the neutral stability

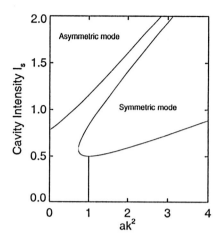

Fig. 7. Neutral stability curves for both the symmetric mode (S-mode) and the asymmetric mode (A-mode) in a self-focusing medium ($\eta = 1, B = 3/2, \theta = 1$).

curve is shown in Fig. 7 for B=3/2, $\theta = 1$. Because the threshold depends only on θ_K, not θ or aK^2 separately, the effect of variations of θ is simply to translate horizontally the curves in Fig. 7. When the S-mode curve meets the vertical axis ($K = 0$), one gets plane-wave optical bistability.

For a self-defocusing medium ($\eta = -1$), once again both the S-mode and the A-mode may become undamped. However, the neutral stability formulas show that the S-mode is always damped if $\theta < \sqrt{3}$, i.e., in the monostable regime. An example of the neutral stability curve is shown in Fig. 8 for B=3/2, $\theta = 1$, where the A-mode first becomes unstable at $(ak_c^2, I_s^c) = (2/3, 2/3)$. Note that Fig. 8 matches onto Fig. 7 if reflected in the vertical axis. This follows from the fact that the thresholds depend on ηaK^2 only. An increase of θ will eventually bring the S-mode curve across the axis to appear in Fig. 8.

Pattern Formation. In self-focusing media the S-mode dominates, so close to threshold, pattern-forming dynamics will be dictated by this mode. This corresponds to the scalar case analyzed in [FSM]. Hexagons dominate close to threshold in a manner broadly similar to that found in the feedback systems discussed above. There is, therefore, no need to discuss this case further.

In the monostable regime, a defocusing medium may become unstable only to the A-mode. Thus, at least close to threshold, pattern-forming dynamics will be governed by this mode. For an A-mode, however, the quadratically nonlinear term that gives rise to the domination of hexagons is absent, allowing a richer variety of spatial structures to form. In other cases where this occurs, rolls tend to dominate in pattern-forming systems, though other patterns may also be observed, such as traveling waves [JMN] and rhombi (squares) [GIM].

The question of pattern selection is addressed both by direct numerical simulation of the model equations (14) and by nonlinear analysis in the vicinity

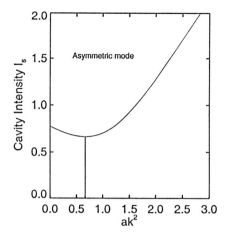

Fig. 8. Neutral stability curve for the asymmetric mode (A-mode) in a self-defocusing medium ($\eta = -1, B = 3/2, \theta = 1$).

of the bifurcation point in [GMW]. Using standard techniques [CH], an amplitude equation is derived for E, which, to the third order, takes the form of a Newell-Whitehead-Segel equation,

$$\tau_0 \partial_t F = \mu \varepsilon F + \xi_0^2 (\partial_x - \frac{i}{2K_c} \partial_y^2)^2 F - \gamma |F|^2 F. \tag{23}$$

Here F is the complex amplitude of the roll, assumed to be slowly varying, as is normal in such problems. For a defocusing medium with $B = 3/2$ and $\theta = 1$, one obtains $\tau_0 = 2, \mu = 2, \xi_0 = 2aK_c$, and $\gamma = 0.611112$. In only one transverse dimension (x), equation (23) reduces to a real Ginzburg-Landau equation, which has homogeneous solutions of the form

$$F = \left(\frac{\mu \varepsilon}{\gamma}\right)^{1/2}. \tag{24}$$

In Fig. 9, we show the amplitude determined by equation (24) versus the numerical results of one-dimensional simulations for B=3/2, $\theta = 1$. Close to threshold ($\varepsilon = 0$) the agreement is excellent, while as expected, the expansion slowly loses validity as ε is increased.

It is worth remarking that the one-dimensional model is applicable also in the time domain, with the diffraction Laplacian reinterpreted as describing dispersion, e.g., in a monomode optical fiber. The latter corresponds to the work of Haelterman and colleagues [HTW].

In two transverse dimensions, the analysis predicts that rolls dominate close to threshold, which is confirmed by simulations of (14) for both quasi-plane wave and super-Gaussian excitation [GMW]. Figures 10 and 11 show a selection of images depicting the near-field intensities of different polarization components for B= 3/2, $\theta = 1$.

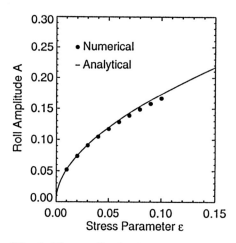

Fig. 9. The amplitude of the roll solution versus relative distance above threshold. The solid dots are obtained by numerical simulation of (14), in one transverse dimension, while the solid line is the solution determined by equation (24).

Figure 10 shows the results of numerical simulations on a periodic domain. These examples are for 10 percent (Fig. 10a,b) and 50 percent (Fig. 10c,d) above threshold. Both show roll-dominated patterns with the calculated transverse wavevector, but consist of patches of rolls with different alignments, separated by *defects* of several kinds. These patterns are metastable, continuing to evolve (anneal) on a rather slow time scale. Figure 10a and b clearly demonstrate the asymmetry between the two senses of circular polarization. Figure 10c and d, on the other hand, are 50 percent above threshold, and pictured in the linear polarization basis, so that Fig. 10d would correspond to imaging through a crossed polarizer.

Figure 11 simulates finite-beam excitation (super-Gaussian profile). As well as being physically more realistic, the finite beam nicely illustrates the effect of the boundary of the excited region. Rolls prefer to align perpendicular to the boundary, but a circular boundary is incompatible with a perfect roll solution. Figure 11a shows, for 10 percent above threshold, a structure that is almost identical to that calculated by Manneville [Man] for fluid convection. In Fig. 11b, we see the stationary solution for 50 percent above threshold with super-Gaussian excitation, which exhibits an intriguing selection of defects. Again, these are associated with the preferred alignment of the rolls normal to the boundary of the excitation region.

For higher excitation, the plane wave polarization instability indicated in Fig. 8 comes into play, and the system breaks up into *polarization domains*. These domains could form the basis of information storage or processing, depending on how high and narrow the domain wall can be made. This is primarily a materials problem, though limited in the present model by the paraxial approximation. The basic mechanism for polarization instability seems likely to survive into the nonparaxial regime.

Fig. 10. (a) Near-field intensity in the right circularly polarized wave at 10 percent above threshold ($B = 3/2, \theta = 1, I_s = 1.1I_s^c$), consisting of rolls of different orientation separated by grain boundaries. Also present are several dislocations. (b) Corresponding left circularly polarized wave, showing the asymmetry between the two components of circular polarization. (c) Near-field intensity in the x-component of the polarization at 50 percent above threshold ($B = 3/2, \theta = 1, I_s = 1.5I_s^c$). (d) Corresponding intensity in the y-component of the polarization, showing rolls punctured by a variety of fascinating defect structures.

3.3 Pattern Formation in a Two-Level Optical Cavity

In this section, we outline the adaptation of the Kerr cavity model [FSM] to the case of a two-level medium in the ring cavity. The atomic detuning Δ and cooperativity C (related to the atomic density), then enter as new control parameters. This enlarges the state space of the Kerr model, allowing for rolls as

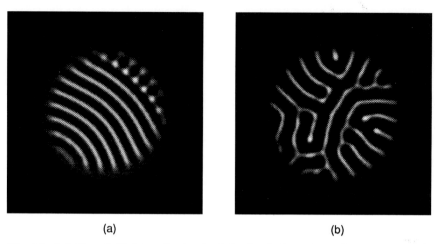

(a) **(b)**

Fig. 11. (a) Near-field intensity in the left circularly polarized wave at 10 percent above threshold ($B = 3/2, \theta = 1, I_s = 1.1I_s^c$) under super-Gaussian excitation. The pattern shows two grains of rolls separated by a grain boundary close to the edge of the domain of excitation. (b) Near-field intensity in the right circularly polarized wave at 50 percent above threshold ($B = 3/2, \theta = 1, I_s = 1.5I_s^c$), again displaying rolls and long-lived defect structures.

well as hexagons and for competition between them [FS]. In certain cases, *localized* solutions are found, and their nature and possible means of configuring them as pixels in an optical memory are described.

If the atomic variables relax fast enough, the two-level ring cavity may be described by a single partial differential equation in the complex field E

$$\dot{E} = -E\left[(1 + i\theta) + \frac{2C(1 - i\Delta)}{1 + \Delta^2 + |E|^2}\right] + E_I + ai\nabla^2 E. \tag{25}$$

This (scalar) equation is the two-level generalization of the Kerr cavity [LL, FSM, LO]. The limits of validity of this equation remain a matter for debate [MGE, TM, FS]. In this author's view, the case of $\theta = O(1)$, $|\Delta| \ll 1$ is regular, so that it is admissible to analyze equation (25) in that limit at least. The results presented are for $\Delta = 0$, within the aforesaid limits, and should be robust enough to justify experimental investigation.

Steady-state homogeneous solutions, E_s, of (25) are known. For suitable values of θ, Δ, C, and $I_s = |E_s|^2$, they describe the S-shaped characteristic familiar from optical bistability (OB). The pattern-forming phenomena described below do not require that I_s be a multivalued function of E_I.

Consider the resonant-excitation case ($\Delta = 0$). From the known features of plane wave ($K = 0$) absorptive OB, we can infer that there is a pattern-forming instability with $K = K_c$, where

$$\theta + aK_c^2 = 0. \tag{26}$$

This again is precisely the condition under which the tilt of the perturbation wave cancels the cavity mistuning. We see that θ must be negative for (26) to be satisfied for real K. The threshold curves for $K \neq K_c$ are always higher than for $K = K_c$, which will thus be dominant [FS].

Nonlinear Analysis. For perturbations with the critical wavevector the unstable component of E is real, while the imaginary part is strongly damped. We can thus regard $Im(E)$ as *slaved* to $Re(E) = R$. We can thus solve its equation algebraically and substitute to obtain (to third order in R)

$$\dot{R} = \left(\frac{2C(I_s - 1)}{(1 + I_s)^2} - 1 \right) R - \frac{(1 + I_s)a^2(\nabla^2 + K_c^2)^2}{1 + I_s + 2C} R$$

$$- \frac{2CI_s(I_s - 3)}{(1 + I_s)^3} R^2 + \frac{2CI_s((1 + I_s)^2 - 8I_s)}{(1 + I_s)^4} R^3. \tag{27}$$

This equation displays all the essential features of the problem, and qualitatively explains all the numerical results to be discussed. Note that (27) is valid only where the spatial spectrum of R is concentrated around $|K| = K_c$.

Threshold occurs where the first two (linear) terms vanish together. The first term is space independent and shows the minimum threshold to obey

$$2C(I_s - 1) = (1 + I_s)^2, \tag{28}$$

as in absorptive OB. The second term determines the spatial properties of R. If we consider a simple roll solution of the form $R = G\cos(K_c x)$, then we find, using (26) and (28), that the amplitude G of the roll has a diffusive response. The diffusion coefficient is positive (i.e., physically reasonable) if θ is negative, which is true by assumption in (27). It is of the same order as θ, which is assumed to be of order unity.

The R^2 term in (27) is responsible for hexagon formation as in the Kerr-slice problem discussed earlier. Here, however, this term can have either sign. For $I_s < 3$, the pattern is a hexagonal array of bright spots (H^+) as in previous cases [AlF, FSM, GIM], but for $I_s > 3$, one obtains π-hexagons (H^-), i.e., a honeycomb pattern [FS]. Such patterns were recently synthesized in a "double-pass" liquid crystal light valve experiment [PRA], and have also been observed in numerical studies of "nascent OB" [TM].

The final, R^3 term, describes a saturation of the amplitude of the pattern, provided this term has the appropriate sign, which we see to be conditional. For roll patterns, the cubic nonlinearity is stabilizing, if

$$8I > (1 + I_s)^2 \ i.e., \ I_s < 3 + 2\sqrt{2}. \tag{29}$$

For intensities $I_s > 3 + 2\sqrt{2}$, the rolls are *subcritical*, and to obtain a steady-state solution to the amplitude equations, we must expand (25) to higher order in E. This case allows for localized, nonperiodic spatial structures, as discussed in the next section.

Isolated States. Subcritical rolls imply a *coexistence* region of the flat solution with a roll solution (which may be unstable). It is precisely in this region that *localized states* are found [TM, SFD]. These states are like one or more peaks of a roll (or hexagon in two dimensions) on a flat background. Figure 12 shows an example for our case of $\Delta = 0$. Such soliton-like states can only exist if the flat background is stable.

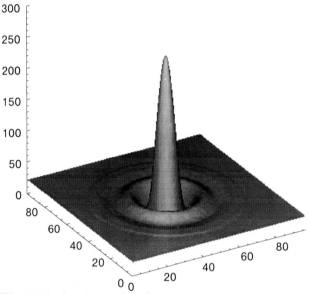

Fig. 12. Isolated state (spot) for a resonant two-level system in a ring cavity. Courtesy of A.J. Scroggie.

In two dimensions, the nonlinear Schrödinger equation does not possess soliton solutions in a self-focusing medium, because any localized state either disperses or collapses to a catastrophic self-focusing singularity [New]. Here there is dissipation and also the nonlinearity saturates, so isolated soliton-like states (let's call them *spots*) may be allowed. Spot solutions are also found numerically in the Kerr cavity model [SFD].[4]

To find applications for these spots, it is necessary to be able to generate, store and erase them. The first is straightforward: a local address pulse will create a spot at the point of address (at least for appropriate parameter ranges). The spot is stationary in principle, but because its location is arbitrary (due to the translational symmetry of (25)), there is no restraint against its wandering away from its original location. Thus some method of fixing the position of the

---

[4] Ed. Note: Spot solutions can be created in 2-D feedback systems with a global feedback (see Chapter 2 and [VK] or in a nonlinear interferometer using semiconductive nonlinear media [AK]).

spots is desirable. In one transverse dimension, a similar problem was overcome by imposing a small constant amplitude modulation on the input pump beam: The "spots" rode up the gradient to the peaks and remained there [McF].

Here, there is another possibility. The basic equation (25) has a Galilean-like "boost" symmetry. Suppose that the input beam has a phase gradient (e.g., it is slightly misaligned). Then this symmetry implies that there are solutions that are transversely *moving* versions of any static solution that corresponds to a flat input phase. It turns out that spots will move *up* a phase gradient. Therefore, if a pattern of phase peaks and troughs is present on the input beam, and a spot is created at any point, it should move to the nearest phase peak and remain there. Because at each peak we may choose to have a spot or not, we thus have a binary optical memory, with each peak a "pixel."

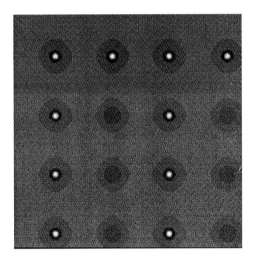

Fig. 13. A 4 x 4 array of pixels with 10 spots induced, so as to store and display the letters "IT." Courtesy of A.J. Scroggie.

Figure 13 shows that this works in simulations [ScF]. If the phase gradients are too steep, single spots can generate neighboring spots, and the field bursts into a pattern filled with spots. Nevertheless, the principle seems to be viable.

The last requirement of a memory is to erase the spots, which can be done, e.g., by locally removing the pump from the pixel. The *capacity* of such a memory is 2^{mn} bits for an m x n array of phase peaks.

4   Conclusion

Passive nonlinear optical systems present a rich variety and great flexibility in relation to spontaneous pattern formation. Feedback systems, both intrinsic and hybrid, have proved very fertile for both theory and experiment, and some of their main features have been summarized in this article.

In discussing mean-field Kerr cavities, it has been shown that the scalar approximation is rigorously valid only for circularly polarized input fields. For linearly polarized input fields and defocusing media, the first instability threshold may involve the generation of depolarized fields. These may appear with a finite transverse wavevector forming patterns. The nature of the instability is such that these patterns are quite different from the hexagons typically seen in scalar models. Boundary effects seem to play a critical role, which adds to the interest and challenge of finding experimental systems that show these features.

Some novel and interesting forms of spontaneous optical pattern formation in a model of a ring cavity containing a two-level system have been discussed. The essentially geometric nature and generality of the tilted-wave mechanism for pattern selection in cavities are rendered especially clear for a purely absorptive nonlinearity. Isolated states have been identified and shown to be configurable as pixels in an all-optical memory.

Studies of optical pattern formation are now reaching the stage where those involved can embark on a fruitful interaction with related fields, such as optical computing, image processing, and optical neural networks.

Acknowledgments The work presented here has been carried out in collaboration with many friends and colleagues, whose contributions are gratefully acknowledged. In particular, Giampaolo D'Alessandro, Francesco Papoff, Gian-Luca Oppo, Mikhail Vorontsov, John Geddes, Andrew Scroggie, and Ewan Wright are thanked for access to source articles. Any errors are, however, the sole responsibility of the author. This work was supported by the EC via ESPRIT grant 7118 (TONICS), by a NATO travel grant, and by SERC GR/G 12665.

References

[Lug]   Lugiato, L.A.: In Progress in Optics. Ed. Wolf, E., **XXI**, North Holland, Amsterdam, 1984, 69.

[Gib]   Gibbs, H.M.: Optical Bistability: Controlling Light with Light. Academic, Orlando, 1985.

[Gbs]   Moloney, J.V., and Gibbs, H.M.: Phys. Rev. Lett. **48** 1982, 467.

[MHG]   Moloney, J.V., Hopf, F.A., and Gibbs, H.M.: Phys. Rev. A **25** 1982, 3442.

[Gry]   Grynberg, G.: Opt. Comm. **66** 1988, 321.

[BVS]   Grynberg, G., Le Bihan, E., Verkerk, P., Simoneau, P., Leite, J.R.R., Bloch, D., LeBoiteux, S., and Ducloy, M.: Opt. Comm. **67** 1988, 363.

[FP]   Firth, W.J., and Paré, C.: Opt. Lett. **13** 1988, 1096.

[PPM]   Petrossian, A., Pinard, M., Maître, A., Courtois, J-Y., and Grynberg, G.: Europhys. Lett. **18** 1992, 690.

[GIM]   Geddes, J.B., Indik, R.A., Moloney, J.V., and Firth, W.J.: Phys. Rev. A (in press).

[BL]   Oppo, G-L., Brambilla, M., and Lugiato, L.A.: Phys. Rev. A **49** 1994, 2028.

[TQG]   Tredicce, J.R., Quel, E.J., Ghazzawi, A.M., Green, C., Pernigo, M.A., Nar-
        ducci, L.M., and Lugiato, L.A.: Phys. Rev. Lett. **62** 1989, 1274.
[LPN]   Lugiato, L.A., Prati, F., Narducci, L.M., and Oppo, G-L., Opt. Comm. **69**
        1989, 387.
[AGR]   Arecchi, F.T., Giacomelli, G., Ramazza, P.L., and Residori, S.: Phys. Rev.
        Lett. **65** 1990, 2531.
[JMN]   Jakobsen, P.K., Moloney, J.V., and Newell, A.C.: Phys. Rev. A **45** 1992, 8129.
[AF]   Abraham, N.B., and Firth, J.W.: J. Opt. Soc. Am. B **7**, 951 (1990).
[Lgi]   Lugiato, L.A.: Phys. Rep. **219** 1992, 293.
[Wei]   Weiss, C.O.: Phys. Rep. **219** 1992, 311.
[Lia]   Special issue on Nonlinear Optical Systems, Chaos, Noise. Ed. Lugiato, L.A.,
        to appear in Chaos, Solitons and Fractals.
[Roy]   Chaos in Optics. Ed. Roy, R., SPIE Proceedings **2039** 1993; Session 5 and op.
        cit.
[AVI]   Akhmanov, S., Vorontsov, M.A., and Ivanov, Y.: JETP Lett. **47** 1988, 707.
[KBT]   Kreuzer, M., Balzer, W., Tschudi, T.: Mol. Cryst. Liq. Cryst. **198** 1990, 231.
[ME]   Macdonald, R., and Eichler, J.H.: Opt. Comm. **89** 1992, 289.
[TBW]   Tamburrini, M., Bonavita, M., Wabnitz, W., and Santamato, E.: Opt. Lett.
        18 1993, 855.
[Hon]   Honda, T.: Opt. Lett. **18** 1993, 598.
[Ban]   Banerjee, P.P.: Nonlinear Optics: Materials, Fundamentals, Applications.
        Hawaii, July 1994. (Paper MP4).
[GVP]   Giusfredi, G., Valley, J.F., Pon, R., Khitrova, G., and Gibbs, H.M.: J. Opt.
        Soc. Am. B **5** 1988, 1181.
[GMG]   Gibbs, H.M., McCall, S.L., Gossard, T.L.C., Passner, A., and Wiegman, W.:
        Appl. Phys. Lett. **35** 1979, 451.
[MSJ]   Miller, D.A.B., Smith, S.D., and Johnston, A.: Appl. Phys. Lett. **35** 1979, 658.
[Frh]   Firth, W.J.: J. Mod. Opt. **37** 1990, 151.
[AF]   D'Alessandro, G., and Firth, W.J.: Phys. Rev. Lett. **66** 1991, 1597.
[AlF]   D'Alessandro, G., and Firth, W.J.: Phys. Rev. A **46** 1992, 537.
[GMP]   Grynberg, G., Maître, A., and Petrossian, A.: Phys. Rev. Lett. **72** 1994, 2379.
[CCL]   Ciliberto, S., Coullet, P., Lega, J., Pampaloni, E., and Perez-Garcia, C.: Phys.
        Rev. Lett. **65** 1990, 2370.
[Pom]   Pomeau, Y.: Physica D **23** 1986, 3.
[SS]   Shtilman, L., and Sivashinsky, G.: Physica D **52** 1991, 477.
[ShS]   Shtilman, L., and Sivashinsky, G.: Can. J. Phys. **68** 1990, 768.
[VFi]   Vorontsov, M.A., Firth, W.J.: Phys. Rev. A **49** 1994, 2891.
[PaA]   Papoff, F., D'Alessandro, G., Oppo, G-L., and Firth, W.J.: Phys. Rev. A **48**
        1993, 537.
[PRR]   Pampaloni, E., Residori, S., Ramazza, P-L., and Arecchi, F.T.: Europhys. Lett.
        25 1994, 587.
[VL]   Vorontsov, M.A., and Larichev, A.V.: Proc. SPIE **1409** 1991, 260.
[TNT]   Thüring, B., Neubecker, R., and Tschudi, T.: Opt. Comm. **102** 1993, 111.
[Vor]   Vorontsov, M.A.: Proc. SPIE **1402** 1991, 116.
[PRA]   Pampaloni, E., Residori, S., and Arecchi, F.T., Europhys. Lett. **24** 1993, 647.
[MMN]   McLaughlin, D.W., Moloney, J.V., and Newell, A.C.: Phys. Rev. Lett. **54** 1985,
        681.
[LL]   Lugiato, L.A., and LeFever, R.: Phys. Rev. Lett. **58** 1987, 2209.
[LO]   Lugiato, L.A., and Oldano, C.: Phys. Rev. A **37** 1988, 3896.

[QAM]   Quarzeddini, A., Adachihara, H. and Moloney, J.V.: Phys. Rev. A **38** 1988, 2005.

[MAI]   Moloney, J.V., Adachihara, H., Indik, R., Lizarraga, C., Northcutt, R., McLaughlin, D.W., and Newell, A.C.: J. Opt. Soc. Am. B **7** 1990, 1039.

[HVR]   Haelterman, M., Vitrant, G., and Reinisch, R.: J. Opt. Soc. Am. B **7** 1990, 1309.

[Vit]   Vitrant, G., Haelterman, M., and Reinisch, R.: J. Opt. Soc. Am. B **7** 1990, 1319.

[FSM]   Firth, W.J., Scroggie, A.J., McDonald, G.S., and Lugiato, L.A.: Phys. Rev. A **46** 1992, R3609.

[HTW]   Haelterman, M., Trillo, S., and Wabnitz, S.: Opt. Comm. **91** 1992, 401.

[GMW]   Geddes, J.B., Moloney, J.V., Wright, E.M., and Firth, W.J.: Opt. Comm. (in press).

[FS]   Firth, W.J., and Scroggie, A.J.: Europhys. Lett. **26** 1994, 521.

[MB]   McCord, A.W., and Ballagh, R.J.: J. Opt. Soc. Am. B **7** 1990, 73.

[CH]   Cross, M.C., and Hohenberg, P.C.: Rev. Mod. Phys. **65** 1993, 851.

[Man]   Manneville, P.: Dissipative Structures and Weak Turbulence. Academic, New York, 1990, 295.

[MGE]   Mandel, P., Georgiou, M., and Erneux, T.: Phys. Rev. A **47** 1993, 4277.

[TM]   Tlidi, M., and Mandel, P.: Special issue on Nonlinear Optical Systems, Chaos, Noise, to appear in Chaos, Solitons and Fractals.

[SFD]   Scroggie, A.J., Firth, W.J., McDonald, G.S., Tlidi, M., LeFever, R., and Lugiato, L.A.: Special issue on Nonlinear Optical Systems, Chaos, Noise, to appear in Chaos, Solitons and Fractals.

[New]   Newell, A.C.: Solitons in Mathematics and Physics. Philadelphia: SIAM, 1987.

[McF]   McDonald, G.S., and Firth, W.J.: J. Opt. Soc. Am. B **7** 1990, 1328.

[ScF]   Scroggie, A.J., and Firth, W.J.: To be published.

[VKS]   Vorontsov, M.A., Koriabin, A.V., Shmalhauzen, V.I.: Controlling Optical Systems. Nauka, Moscow, 1988.

[LMW]   Leite, R.C., Moore, R.S., Whinnery, J.R.: Appl. Phys. Lett. **5** 1964, 141.

[AKM]   Akhmanov, S.A., Kzindacz, D.P., Migulin, A.W., Suchorukow, A.P., Khokhlov, R.V.: IEEE J. Quantum Elect. **QE-4** 1968, 568.

[FV]   Firth, W.J., and Vorontsov, M.A.: Opt. Comm., 1993, 1841.

[AVL]   Akhmanov, S.A., Vorontsov, M.A., Ivanov, V.Yu., Larichev, A.V., and Zheleznykh, N.I.: J. Opt. Soc. Am. B, 9 **1**, Jan. 1992, 78.

[VK]   Vorontsov, M.A., and Kobzev, E.N.: Proc. Soc. Photo-Opt. Instrum. Eng. SPIE **1402** 1991, 1965.

[VMi]   Vorontsov, M.A.: Sov. J. Quantum Electron. **20** 1993, 319.

[ALV]   Arecchi, F.T., Larichev, A.V., Vorontsov, M.A.: Opt. Comm. **105** 1994, 297.

[AK]   Akhmanov, S.A., Khokhlov, R.V.: On Trigger Properties of Nonlinear Wave-Guiding Systems. Izv. Vyssh. Uchebn. Zaved. Radiofiz. **5** 1962, 742.

4 Spatio-Temporal Instability Threshold Characteristics in Two-Level Atom Devices

M. Le Berre, E. Ressayre, and A. Tallet

Laboratoire de Photophysique du C.N.R.S., Université de Paris-Sud, 91405 Orsay, France

Introduction

Multistable nonlinear optical cavities and lasers provide an excellent vehicle to study examples of spatio-temporal chaos. As a matter of fact, instabilities in optics have slowly emerged as a whole domain, partly because the earlier studies dealt with solid-state lasers, which despite their economical importance, have proved very difficult to modelize [Cas]. Later, quantitative modelization for gas lasers progressed [WAb, Are, HLB], moreover for passive systems [Ike] a lot of work has been done since Ikeda's prediction [Ida, IkD] of instabilities in a ring cavity. New transverse structures, such as extra rings or multispots set on a ring in the far-field pattern [Gry, GBV, PHe, PeP, KMG] or [SeM, SML, Oro, Gra], new frequencies, or both spatial and temporal effects [PeP, Khi, Tai, Giu, AVI, Vor] were observed from incident cw or pulsed Gaussian beams. Since the pioneer work of Moloney [MMN, MAM], theoretical works dealing with dynamics of patterns mostly concern optical media with third-order nonlinearity (Kerr media) [Fir, Ale, Cha, Ged, FiP, GrP, VF, MAI]. For two-level media in optical cavities, the numerical and analytical investigations are so heavy that up to now the various approaches to this problem have consisted of either reducing the number of variables [LeB] or performing a perturbation treatment as we do here. One of the most ingenious treatments is the uniform-field model developed by Lugiato and colleagues [LOT], which has the great advantage of getting rid of the propagation inside the medium, but concerns situations in which only one longitudinal mode of the empty cavity contributes to the resulting spatio-temporal pattern.

Here we consider passive two-level atomic media pumped by continuous light beams in situations typically outside the uniform field conditions. Three experiments are described and analyzed. In the first experiment, realized by H.M. Gibbs's group [Tai, Giu] a unique laser light is sent through a sodium cell and reflected back to the medium via a spherical mirror, i.e., a half-cavity device. The second experiment concerns the waveguide instabilities observed by Ségard and Macke [SeM, SML] in a Fabry-Pérot interferometer filled with $HC^{15}N$ molecules, i.e., the medium is located between two plane mirrors forming a cavity. There is no mirror in the third experiment [Khi] relating a Rayleigh self-oscillation observed in sodium vapor lighted from both sides by two counter-propagating laser beams. These three different configurations are often seen as *feedback devices*: In the first two experiments, an external feedback is provided via one or two mirrors, while in the third experiment, there is only the feedback self-generated by the medium. In all cases, counter-propagating pump waves may lead the two-level system to emit new frequencies along directions close to the optical axis defined by the direction of the pumps.

In each of these three cases, our aim is to propose the simplest model capable of reproducing the first Hopf bifurcation, which generates the observed self-oscillation. To investigate the threshold, we perform a linear stability analysis of the Maxwell-Bloch equations. A linear analysis is a tool obviously much more restrictive than a complete numerical investigation, nevertheless, this method allows us to study the role of various mechanisms involved in the bifurcation process.

We have especially focused on the role of the *standing wave*[1] and *transverse grating* effects in such devices, on the influence of the polarization dephasing time T_2, of the *detuning* δ between the laser and atomic angular frequencies, of the *atomic concentration* and of the *delay effects*.

The semianalytical character of the linear analysis allows us to investigate the properties of the nonlinear medium in the vicinity of the self-oscillation threshold. For such problems, it is crucial to understand the origin of the new frequencies. Up to now, self-oscillations were often interpreted in terms of *gain* concept, with the underlying idea that the system preferably emits in the frequency range where the nonlinear medium gain has a strong amplification. Before any interpretation, it seems necessary to precisely define these notions, especially because some confusion may exist between the gain of the full device and the gain of the isolated pumped nonlinear medium.

Spatio-temporal instability arises in an optical device when, for a given value of a control parameter, a new field $\epsilon_{out} \exp(i\omega t + \mathbf{kx})$, with spatial wavevector $\mathbf{k}$ and frequency ω, appears at the output of the device, while the input amplitude ϵ_{in} of this component is equal to zero. Then the onset of instability can be defined by the relation

$$g_{th}^c(\mathbf{k}, \omega) = |\epsilon_{out}/\epsilon_{in}| = \infty, \tag{1}$$

where the ratio $\epsilon_{out}/\epsilon_{in}$ is the gain of the full device formed by the pumped nonlinear medium, plus the mirrors, if there are any.[2] If there is no cavity, the quantity in (1) exactly displays the gain properties of the isolated pumped nonlinear medium, also called an "intrinsic system". But if the medium is placed inside a cavity or a half-cavity, *the dressed cavity gain* $g^c(\mathbf{k}, \omega)$ is different from the *intrinsic system gain* $g^i(\mathbf{k}, \omega)$ For calculation of the dressed cavity gain, we

---

[1] Ed. Note: Another term, dissipative structures, is also used. Also, the term *standing wave* includes other effects, as explained in Sect. 2.

[2] Ed. Note: The parameter $g_{th}^c(\mathbf{k}, \omega)$ is characterized by the spectral transverse function of a device, however, in atomic physics, the term "gain" is more commonly used.

must take into account not only the propagation through the medium, but also the boundary conditions due to the presence of the mirrors.

Keeping in mind these two distinct gain functions, it is tempting to predict that an intrinsic system will emit a new field in the frequency domain where the intrinsic gain $g^i(\mathbf{k}, \omega)$ is large; and that an optical cavity will obey the pleasant rule of thumb [SML], which states that self-oscillation occurs when one of the cavity modes is resonant with one of the largest positive gain frequency domains. What are the positive gain frequency domains?

Let us recall that in the presence of a unique strong pump beam of frequency ω_f, the transmission spectrum [Mol, Har] of a copropagating weak probe through a two-level medium displays "positive gain" (i.e., g^i larger than unity, or $\ln(g^i)$ positive) in two distinct frequency domains. These domains are labeled the "Rayleigh" and "Rabi" domains [Har], the former with a width of the order T_1^{-1} around the laser frequency ω_f, corresponds to an elastic process; the latter gain, which results from a three-photon process due to the atomic nutation, is positive around the frequency ω_R shifted from the laser frequency ω_f by approximately the generalized Rabi frequency $\sqrt{\Omega_0^2 + \delta^2}$.

In the case of two counter-propagating pump beams with equal input intensity Ω_0^2 (in terms of the Rabi frequency), the intensity oscillates along the cell between zero and $4\Omega_0^2$.[3] Therefore, the Rabi gain is expected to be positive in the frequency range $\{\delta, \sqrt{4\Omega_0^2 + \delta^2}\}$ if the atomic nutation is effective, i.e., for input fields of amplitude ω_0 of magnitude of order (or larger than) the detuning δ. This property is confirmed by our study.

The onset of self-oscillation in standing-wave devices reveals itself as a complex process. In a device without a cavity, the threshold gain curve displays infinite value for frequencies $\omega = \omega_f \pm \beta_H$, which are most often outside the Rayleigh or Rabi frequency ranges (where the gain may be either large or small depending on the values of the parameters).

In optical cavities, the rule of thumb has no a priori reason to be valid, because the threshold condition in (1) does not necessarily coincide with large gain g^i. On the contrary, here it is shown that for the Fabry-Pérot device and elsewhere for the ring cavity device [Pat] relation (1) is satisfied for *small intrinsic gain* ($g^i \approx 1$) only. Finally, the threshold relation for the half-cavity does not involve the gain function, it only concerns the reflectivity properties of the medium, which explains why the feedback mirror device behaves like a resonator, with the nonlinear medium acting as a second mirror.

The Maxwell-Bloch equations describing the evolution of the electric fields and atomic variables are given in Sect. 1, together with an outline of the linear stability analysis for counter-propagating beams. The three experiments selected to perform our investigation are very interesting for the following reasons: The first experiment (Sect. 2) is very exciting, because of the various spatio-temporal patterns exhibited, the large domain of parameters being investigated, and the period being so invariant, all of which cause *the experiment to act efficiently as*

---

[3] Author's Note: Ω is the Rabi frequency, and with an appropriate scaling, one can define the intensity as Ω_0^2.

a guide to select the right model. The second experiment represents a sort of "ideal" case from a theoretical point of view, because there is only one transverse mode in the waveguide and the atoms are free of Doppler effect (Sect. 3). Finally, the third experiment reports a Rayleigh gain lasing in contradiction to the contemporary theoretical predictions (Sect. 4), which has fruitfully served to question the theory.

1 Linear Stability Analysis of Stationary Solutions

Let E be the amplitude of the positive frequency part of the linearly polarized electric field E, which slowly varies with time,

$$E(r, z, t) = \frac{1}{2} \left[\mathrm{E}(r, z, t) e^{-i\omega t} + \text{c.c.} \right] \tag{2}$$

and P the amplitude of the atomic polarization

$$P(r, z, t) = \frac{1}{2} \left[\mathrm{P}(r, z, t) e^{-i\omega t} + \text{c.c.} \right]. \tag{3}$$

Setting $\epsilon = (\mu/\hbar)\mathrm{E}$, where μ is the dipole moment and $\Delta = T_2(\omega_{ab} - \omega_L)$, the Bloch equations for the evolution of the atomic population $w(r, z, t)$ and the polarization amplitude $\mathrm{P}(r, z, t)$ are

$$\frac{\partial w}{\partial t} = \frac{i}{2}(\epsilon \mathrm{P}^* - \epsilon^* \mathrm{P}) - \frac{1}{T_1}(w + 1) \tag{4}$$

$$\frac{\partial \mathrm{P}}{\partial t} = -\frac{1}{T_2}(1 + i\Delta)\mathrm{P} - i\epsilon w. \tag{5}$$

The electric field $\epsilon(r, z, t)$ is a sum of a forward field and a backward field

$$\epsilon(r, z, t) = \epsilon_F e^{+ik_z z} + \epsilon_B e^{-ik_z z}. \tag{6}$$

The amplitudes $\epsilon_{F,B}$ are supposed to slowly vary with respect to z (slowly varying envelope) approximation [RAF], $|\frac{\partial}{\partial z}\epsilon_{F,B}| \ll k_z |\epsilon_{F,B}|$), they obey the propagation equations deduced from the Maxwell equations

$$\left(-\frac{i}{2k}\nabla_\perp^2 + \frac{1}{c}\frac{\partial}{\partial t} \pm \frac{\partial}{\partial z} \right) \epsilon_{F,B} = \frac{i\alpha_0}{2T_2} \int_{\delta z} e^{\pm ik_z z} \mathrm{P} dz, \tag{7}$$

where $\nabla_\perp^2$ is the transverse Laplacian, and α_0 is the line-center absorption coefficient. Equation (7) is valid if the atomic response acts as a perturbation, because the average on the right-hand side singles out the portion of the nonlinear polarization that propagates with the same k vectors as the electric fields. Integration over the length δz is indeed a linear filtering [Pap], which simply corresponds in the k-space to multiply the k-components of the polarization by the function $\frac{\sin(k_z \delta z)}{k_z \delta z}$ of half-width $\Delta k_z \approx 1/\delta z$. In the case of two counter-propagating beams, the slowly varying condition requires that the wave-number spectrum of the electric field have two quasi-monochromatic components around $\pm k_z$, of width Δk_z much smaller than k_z, that requires $\delta z = 1/\Delta k_z$ much larger than $1/k_z$. Generally the appropriate choice of the integration length is the wavelength of the light, $\delta z = 2\pi/k_z$, which seems never to be questioned [McC].

Defining the total stationary intensity $I = \frac{|\epsilon|^2 T_1 T_2}{(1+\Delta^2)}$ scaled to the off-resonance saturation one, the stationary solutions of (4,5)

$$w_s = -\frac{1}{1+I}, \tag{8}$$

$$P_s = -i\frac{1-i\Delta}{1+\Delta^2}\epsilon w_s T_2 \tag{9}$$

contain the expression $\epsilon\epsilon^*$, which is longitudinally modulated via the term $e^{2ik_z z}$, so that the population and the polarization amplitude display a longitudinal grating. These can be expanded in Fourier series

$$w = \sum_n w_n e^{2ink_z z}, \tag{10}$$

$$P = \sum_n P_{2n+1} e^{i(2n+1)k_z z}. \tag{11}$$

Let us return briefly to the hypothesis of the slowly varying envelope. If the integration over the small distance δz were not performed, the right-hand side of (7) would be proportional to $\sum_n e^{ik_z 2nz} P_{2n+1}$, which contains source terms with all odd harmonics of the pump wavevectors. After the integration over δz, this series becomes

$$\sum_n e^{ik_z(2nz+n\delta z)} \frac{\sin(nk_z\delta z)}{nk_z\delta z} P_{2n+1} .$$

The choice of $\delta z = 2\pi/k_z$ indeed cancels all terms, except the low-frequency term corresponding to $n = 0$, which is consistent with the slowly varying envelope approximation. Nevertheless, this procedure appears at this stage as an artifact. To justify that the harmonics play no role, it would be better to prove that the successive P_{2n+1} for $n > 0$ can be neglected. We have proved [BRZ] that in the limit of small intensity I (scaled to the off-resonance saturation one) the nth term of the series is proportional to I^n, then it is justifiable to keep the lowest order components of the series. But in the limit of large intensity (scaled to the saturation one), all the terms of the series are of equal importance, which brings into question the slowly varying envelope approximation. In the following, we suppose that the slowly varying envelope approximation is valid, because the threshold intensity for the first Hopf bifurcation is generally small.

The propagation equations for the stationary field amplitudes $\epsilon_{F,B}^s$ are

$$\frac{d\epsilon_F^s}{dz} = \frac{\alpha_0}{2}\Phi_3(w_0\epsilon_F^s + w_1\epsilon_B^s) \tag{12}$$

with

$$\Phi_3 = (1+i\Delta)^{-1} \tag{13}$$

and a similar equation for the backward amplitude, where
$z \to -z, F \to B, w_1 \to w_{-1}$ deduced from w_1 by changing $F \to B$.

The linear stability analysis is a first-order perturbative treatment, which can predict the values of the parameters, where the stationary solutions become unstable ("threshold" of instability), and to predict the new frequencies $\pm\beta$ and transverse wave numbers k that are spontaneously created. Near the bifurcation, we suppose exponential deviations $\delta\mathbf{X} = (\delta W, \delta P, \delta\epsilon)$ from the stationary solutions (W_s, P_s, ϵ_s),

$$\delta\mathbf{X} = e^{\lambda t}\delta\mathbf{X}_+ + e^{\lambda^* t}\delta\mathbf{X}_- \tag{14}$$

with $\lambda = \alpha + i\beta$, as originally treated in such systems by Bar-Joseph and Silberberg [BrJ], the only difference being that the small components $\delta\mathbf{X}_\pm$ have a transverse structure that can be generally expanded as

$$\delta\epsilon_F(\mathbf{x}, z, t) = \sum_{\lambda_k} \phi_k(\mathbf{x}, z)e^{\lambda_k t}. \tag{15}$$

The threshold of instability is defined by the following procedure: all parameters are kept constant, except one, called the "control parameter" (for example, the input intensity or the frequency of the incident light), which increases. Under threshold, the real part of all exponents in (15) are negative; at threshold, one of the exponents is such that $\mathrm{Re}\{\lambda_k\} = 0$ defines the transverse mode ϕ_k, which first destabilizes with the appearance of a new pulsation $\beta = \mathrm{Im}\{\lambda_k\}$.

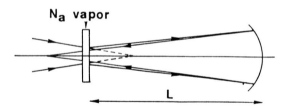

Fig. 1. Experimental setup of the feedback mirror device.

If ϕ_k is a Fourier mode written $e^{i\mathbf{kr}}$, the far-field intensity, which is the square modulus of the Fourier transform of $\epsilon_F + \delta\epsilon_F$, exhibits a spot in the direction of the unstable $\mathbf{k}$-vector; if it is a Laguerre-Gaussian mode, written as $f_{n,m}(r)e^{im\varsigma}$, the forward intensity at the cell exit is of the type $I_f = a + b\cos(\beta t - m\varsigma)$, which appears as m spots turning around the axis with period $2\pi/\beta$, as in Fig. 1. A priori, any expansion is adapted to describe patterns like those displayed in Fig. 2 for the feedback mirror experiment.

However, the analytical treatment is generally untractable if the transverse structure of the pump waves is also taken into account, consequently, we have used the plane-wave model for the pumps [BrJ, LBR, Ber, MLe, LBT]. Within this approximation, the linear stability analysis of the three devices predict far-field patterns with cylindrical symmetry, either unstable rings surrounding the pump beams or self-oscillation collinear to the pumps. The perturbation is expanded here on the basis of the spatial Fourier modes

$$\phi_k(\mathbf{x}) = \epsilon_s(f_{\lambda_k}(\mathbf{k}, z)e^{i\mathbf{kx}} + f_{\lambda_k}(-\mathbf{k}, z)e^{-i\mathbf{kx}}). \tag{16}$$

Equations for $\delta\epsilon_{F,B}$ are derived by linearization of (4,5)

$$\left(-\frac{i}{2k_z}\nabla_\perp^2 + \frac{1}{c}\frac{\partial}{\partial t} + \frac{\partial}{\partial z}\right)\delta\epsilon_F =$$
$$\frac{\alpha_0}{2}\Phi_1(w_0\delta\epsilon_F + w_1\delta\epsilon_B + \epsilon_F\delta w_0 + \epsilon_B\delta w_1), \tag{17}$$

$$\left(+\frac{i}{2k_z}\nabla_\perp^2 + \frac{1}{c}\frac{\partial}{\partial t} + \frac{\partial}{\partial z}\right)\delta\epsilon_F^* =$$
$$\frac{\alpha_0}{2}\Phi_2(w_0\delta\epsilon_F^* + w_{-1}\delta\epsilon_B^* + \epsilon_F^*\delta w_0 + \epsilon_B^*\delta w_1) \tag{18}$$

with

$$\Phi_1 = (1 + i\Delta + \lambda T_2)^{-1} \tag{19}$$
$$\Phi_2^* = (1 - i\Delta + \lambda T_2)^{-1}. \tag{20}$$

Equations for the backward perturbation fields are deduced by the rule: $z \rightarrow -z, F \leftrightarrow B, w_1 \leftrightarrow w_{-1}$. The small deviation δw_n of the population Fourier components is derived in [LBR]. The four-dimensional vector

$$\mathbf{V}(k) = (f_{\lambda_k}(\mathbf{k}, z), b_{\lambda_k^*}^*(-\mathbf{k}, z), b_{\lambda_k}(\mathbf{k}, z), f_{\lambda_k^*}^*(-\mathbf{k}, z)) \tag{21}$$

obeys the propagation matrix equation

$$\frac{dV(k)}{dz} = \begin{pmatrix} h & g & s & t \\ -g' & -h' & t' & s' \\ -s & -t & -h & -g \\ -t' & -s' & -g' & -h' \end{pmatrix} V(k), \tag{22}$$

where the matrix elements $h, g, s, t, \ldots$ depend on ϵ_F and ϵ_B [LBR]. Moreover, they depend on the transverse $\mathbf{k}$ vector via the optical phase shift $\theta_D = \frac{\mathbf{k}^2 l}{2k_z}$, which appears in h, h' only. This dependence upon the modulus of the $\mathbf{k}$ vector simply reflects the cylindrical symmetry of the nonlinear medium. The propagation inside the cell of length l reduces to coupled linear equations (22), that can easily be solved by using the Laplace transformation under the further hypothesis of *negligible absorption in the pump fields*, (i.e., $\alpha_0 l/(1 + \Delta^2) \ll 1$ or input intensity, scaled to the saturation one, which is much larger than $\alpha_0 l$), as supposed in Sects. 2 and 3. In Sect. 4, the depletion of the pump is introduced.

Within the no pump-depletion hypothesis, the propagation through the nonlinear medium of the forward two-components vector

$\mathbf{F}(\mathbf{k}, z) = (f_{\lambda_k}(\mathbf{k}, z), f^*_{\lambda_k}(-\mathbf{k}, z))$ and the similar backward one $\mathbf{B}(\mathbf{k}, z)$, is described with the help of two vectorial equations

$$\mathbf{F}(\mathbf{k}, l) + \mathbf{B}(\mathbf{k}, l) = \mathbf{A}_3 \mathbf{F}(\mathbf{k}, 0) + \mathbf{A}_1 \mathbf{B}(\mathbf{k}, 0),$$
$$\mathbf{F}(\mathbf{k}, l) - \mathbf{B}(\mathbf{k}, l) = \mathbf{A}_4 \mathbf{F}(\mathbf{k}, 0) - \mathbf{A}_2 \mathbf{B}(\mathbf{k}, 0), \tag{23}$$

where the 2 x 2 matrices $\mathbf{A}_i$ $(i = 1, 4)$ are given in [LBR]. The forward amplitude vector $\mathbf{F}(\mathbf{k}, l)$ at the cell exit, can then be written as a result of both gain and reflectivity processes,

$$\mathbf{F}(\mathbf{k}, l) = \mathbf{G} \mathbf{F}(\mathbf{k}, 0) + \mathbf{R} \mathbf{B}(\mathbf{k}, l) \tag{24}$$

via the 2 x 2 matrices $\mathbf{G}$ and $\mathbf{R}$

$$\mathbf{G} = 2(\mathbf{A}_1 + \mathbf{A}_2)^{-1}, \tag{25}$$
$$\mathbf{R} = (\mathbf{A}_1 + \mathbf{A}_2)^{-1}(\mathbf{A}_4 - \mathbf{A}_3), \tag{26}$$

which clearly can be interpreted in terms of *gain* and *reflectivity* of the nonlinear medium. This treatment for light beam propagation inside a two-level atom cell is common to any counter-propagating beam device. The peculiarity of a given experiment lies only in the *boundary conditions*.

2 Feedback Mirror Experiment

2.1 Experimental Results

In the feedback mirror experiment [Tai, Giu], a single forward light beam is sent forward through a sodium vapor cell and reflected back to the cell via a concave mirror located at a distance L, i.e., one can say that the cell is placed in front of a "half cavity" (Fig. 1).

As the power P_{in} of the laser increases, the observations report first the apparition of a ring around the central spot in the far-field profiles, at the same frequency ν_L as the laser. This new pattern can be interpreted as a spatial instability, it is sometimes called "static instability." Then a lasing effect is seen from the cell, with, in most cases, two invariable new frequencies $\nu_L \pm c/4L$, associated with various transverse structures. The far-field profiles detected forward and backward are either cylindrically symmetric with the frequency shift $c/4L$ observed in the central spot or in the ring, or else there is a symmetry breaking, with one, two, or three spots, as shown in Fig. 2 (labeled, respectively P1D, P2D, and P3D), turning around the central spot at the frequency $c/4L$. An example of the scenarios reported by Giusfredi in the plane (P_{in}, ν_L) is shown in Fig. 3.

The experiment can be qualified as "global," in the sense that the authors have reported the precise characteristics of the successive bifurcations (static and dynamic) in the space of parameters (P_{in}, ν_L), for two different atomic concentrations and a large domain of other parameters for which the same invariant frequency could be observed. Argon pressure in the cell varies from 0.3 Torr to 22 Torr (T_2/T_1 varies from 0.1 to 1.5), atomic concentration was between

10^{12} and 10^{13} at/cm^3, the transverse polarization of the input beam was linear, circular, or mixed, and finally, the distance cell-mirror varies by a factor of 5 ($4L/cT_1$ varies from 0.4 to 2), and the cell exit was always located approximately at the lens focus.

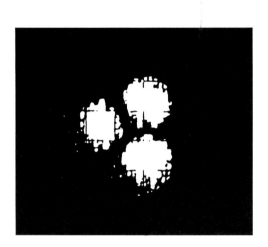

Fig. 2. Stroboscopic view of P3D oscillation (from Giusfredi [Giu]).

The invariance of the generated new frequency, at the value $\nu_L \pm c/4L$, for such a large range of parameters has led the authors to propose an interpretation of the feedback mirror device in terms of *phase-conjugate resonator*. This point will be discussed later.

2.2 Linear Analysis

The first step of the linear analysis concerns the propagation of the fields inside the cell. It was treated in Sect. 1 (23-25). Let us discuss shortly the validity of the plane-wave model for the pump. This model assumes that the interaction between the nonlinear medium and the field is homogeneous in the transverse direction, which may be correct in the central part of the beam. First, it might be violated if the pump beam is distorted when propagating through the cell. But in the experiment, the cell length is about ten times smaller than the Rayleigh length $z_D = 1/2kw_0^2$ (w_0 is the input beam radius), therefore, the diffraction of the pump inside the cell is negligible. Second, if the plane-wave pump model predicts an instability with transverse wavevector $\mathbf{k}$, the result can be compared with an experimental observation if the pump section encloses at least several transverse wavelengths or if $|\mathbf{k}|w_0$ is larger than 2π. Consequently, *the plane-wave pump beam model is reasonable for off-axis instability*.

The second step is the boundary condition for the perturbation, which describes the propagation of the perturbation in the free space between the cell

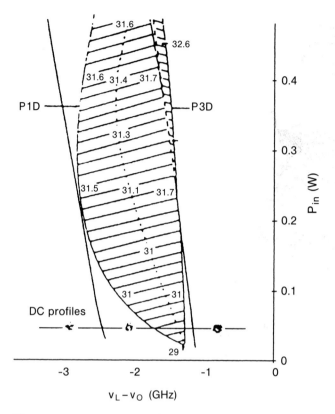

Fig. 3. Instability diagram (from Giusfredi [Giu]).

exit and the mirror. The free-space diffraction is actually treated with a new approximation. Because the cell is located near the focus of the lens, one has $\delta\epsilon_B(\mathbf{x}, l, t) = -\delta\epsilon_F(-\mathbf{x}, l, t - d) \exp(iar^2)$, which generally means that the free-space propagation couples all the k vectors, which would lead to intractable calculations. In this experiment, the phase factor ar^2 is actually irrelevant for r smaller than $w_0/2$, i.e., on the main part of the Gaussian [LBR]. It follows that any forward field $\delta\epsilon_F(\mathbf{x}, l)$ returns to the cell approximately as $\delta\epsilon_B(\mathbf{x}, l) = -\delta\epsilon_F(-\mathbf{x}, l)$, or else only the $+\mathbf{k}$ and $-\mathbf{k}$ vectors are coupled. This approximation allows us to simplify the boundary condition in order to use easily the Fourier modes expansion of the field, one has

$$\mathbf{B}(\mathbf{k}, l) = -\mathbf{F}(-\mathbf{k}, l)e^{-\lambda d}, \qquad (27)$$

with $d = 2L/c$.

Finally the boundary for an instability of period $T = 2\pi/\beta$ follows from the relation

$$\mathrm{Det}(\mathbf{I} - e^{-2\lambda d}\mathbf{R}^2) = 0. \qquad (28)$$

Note that while the boundary condition couples the $+\mathbf{k}$ and $-\mathbf{k}$ vectors, the $\mathbf{R}$ matrix elements in the threshold relation (28) depend on the modulus of $\mathbf{k}$ only. Therefore, the linear analysis cannot predict discrete spots on a ring in the far-field pattern, but rather a pattern with cylindrical symmetry, like rings.

The role of the longitudinal grating of the times T_1, T_2, d, and of the absorption coefficient on the threshold characteristics is studied in detail in [LBR]. The main conclusions are summarized in the following subsection.

2.3 Static and Dynamical Thresholds

Typical static (dashed lines) and dynamic (solid lines) boundaries are reported in Fig. 4, in the parameter space (I, Δ), where I is the input intensity scaled to the on-resonance saturation one, and $\Delta = \delta T_2$ is the scaled detuning for $\alpha_0 l = 5000$, which corresponds to a concentration of 1.25×10^{13} at/cm^3, and $T_2 = T_1 = 2L/c = 16$ ns, a situation close to the observations reported in Fig. 3. The period of the oscillation is written along the curve in units of d, the number in parenthesis is the threshold diffraction term $\theta_D = k l \theta^2/2$, where $\theta = |\mathbf{k}|/k$ is the half-cone emission angle of the self-oscillation.

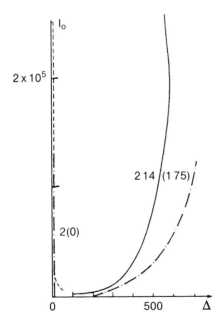

Fig. 4. Static (dotted-dashed line) and dynamic (solid line) boundaries in parameter space (I_0, Δ) for $\alpha_0 l = 5000, T_1 = T_2 = 2L/c$. The system is statically or dynamically unstable above the boundaries.

Let us first consider the *static instability boundary*. The threshold curve strongly depends on the diffraction term θ_D, as illustrated in Fig. 5, for positive (defocusing case, in solid line) and negative (focusing case, in dashed line) detunings. This strong angular selectivity seems in qualitative agreement with the observation of a unique ring in the far-field. Quantitatively, the agreement is also good because the instability threshold in Fig. 4 occurs for $\theta_D = 1.2$, which corresponds to a predicted half-angle $\theta = 5$ mrad. This result appears to be compatible with the experimental observation of a conical shell clearly outside the Gaussian incoming pump beam, which has an aperture of about 2 mrad.

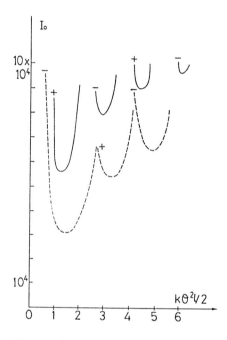

Fig. 5. Threshold intensity versus $\theta_d = kl\theta^2/2$, for the static instability. $\Delta = +600$ (dashed line), $\Delta = -600$ (solid line), other parameters as in Fig. 4.

The external ring of the static profile has only a small percentage of the total intensity, consequently, approximating the distorted pump beam by a plane wave appears to be valid when looking for the Hopf bifurcation above the static threshold.

The agreement between theory and experiment is very good concerning *the period of the dynamical instability*, which is *close to 2d all along the dynamical frontier*. This is the most important result of our study. Indeed, a lot of efforts have previously failed to reproduce this famous 2d period. This point will be discussed in Sect. 2.

As for the extent of the unstable domains, while the left boundaries in Fig. 4 correspond to detunings clearly smaller than the ones in Fig. 3, which is a failure of the model, one must notice that the large detuning part of the theoretical boundaries (right branch) in the (P_{in}, ν_L) diagram are in good agreement with the corresponding experimental ones (which are the left frontiers in Fig. 3).

A striking property concerning the boundary is that the coefficients g, s, t in (22) that govern the multiwave mixing strength are nearly constant along the thresholds. More precisely, numerical values of the coefficient g were investigated. *Along the static boundary*, we find that the nonlinearity

$$|g^{st}| = \alpha l \Delta I_0 (1 + 4 I_0)^{-3/2} \approx 0.6 \qquad (29)$$

has a value close to the well-known threshold for phase conjugation in Kerr media [YaP]. Along the dynamical boundary, the expressions for g, s, t are much more tricky, however, in the right part of Fig. 4, one has $g \approx -0.5i, s \approx -0.8i$. Let us simplify the description of the frontiers by using the same expression g^{st} for quantifying the nonlinearity along the two curves, and let us call g^{st} the "nonlinearity coefficient." In fact, one finds that g^{st} *is practically equal to unity all along the right boundary of Fig. 4*, in which it slowly decreases from 1.2 to 0.8 as $I_0 = \mathcal{I}_0/(1 + \delta^2)$ increases from 0.1 to 1, as reported in the solid line in Fig. 6. The dashed line is the experimental value deduced from the left part of the curve in Fig. 3 (large detunings). The agreement is excellent.

We have also drawn the instability boundaries for higher concentrations of sodium vapor ($\alpha_0 l = 40,000$ and 70,000). It turns out that the above relation (29) persists. More generally, one finds that the dynamic boundary for large detuning depends slightly upon the delay $d = 2L/c$ and is well reproduced by the relation

$$g^{st} = \alpha l \Delta I_0 w_0^3 = \text{constant}, \qquad (30)$$

which also predicts the boundary curvature in agreement with the experiment. Therefore, *it is possible to quantify the nonlinearity along the marginal stability curve by using g^{st} as a scaling factor.*

2.4 Role of the Longitudinal Grating

The previous threshold boundaries were given with the longitudinal grating taken into account. If it is neglected, the instability domain versus the detuning strongly shrinks, which is illustrated by the increasing of the nonlinearity coefficient g^{st}, which becomes about 5. The period is still close to $2d$, but the extension of the domain disagrees with the experimental one. So this result seems to imply that in the experiment the longitudinal grating has not been washed out by the atomic motion.

2.5 Discussion of the Phase Conjugation Effects

Let us compare our results with the threshold for self-oscillation in a phase conjugate resonator, as predicted by Yariv and Pepper [YaP]. In their original description of the phase conjugation induced by four-wave mixing, the component $f_{\lambda_k}(\mathbf{k}, z)$ is supposed to be coupled with $b^*_{\lambda_k}(-\mathbf{k}, z)$ only (pure phase

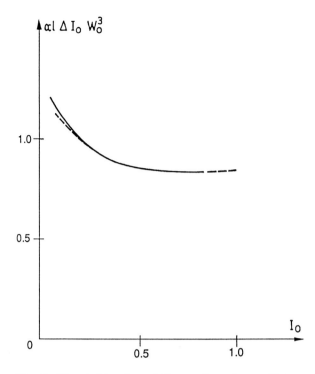

Fig. 6. Threshold value of the nonlinearity coefficient g^{st} along the boundary. The theoretical value, shown in the solid line, is calculated from Fig. 4, the experimental values, shown in the dashed line, are deduced from Fig. 3, both are for the high detuning branch.

conjugation). The threshold is predicted for a coupling coefficient $g = \pi/4$. We were surprised to find similar threshold conditions with our model, including much more complex coupling processes (with other components $b_{\lambda_k}(\mathbf{k}, z)$ and $f^*_{\lambda_k^*}(-\mathbf{k}, z)$). But does the feedback mirror actually behave as a phase-conjugate resonator?

The advantage of the previously discussed semianalytical analysis upon a purely numerical treatment, will now be addressed. At the exit of the cell, the matrix elements of the reflectivity matrix $\mathbf{R}$ in equation (26) can be calculated all along the instability boundaries. Two limiting cases can be simply interpreted: If $\mathbf{R}$ is equal to the identity matrix, the cell behaves like an ordinary mirror; but if $\mathbf{R}$ is off-diagonal and Hermitian, the medium acts as a phase conjugate mirror. *Actually the only simple behavior concerns the left dynamic frontier* in Fig. 4, where the cell indeed behaves approximately like an ordinary mirror. On the right frontier, the forward and backward weak beams are connected by the relation

$$\delta E_f \approx a\delta E_b + b\delta E_b^*, \qquad (31)$$

with $|a| \approx |b|$. The relation (31) means that the nonlinear medium acts as a

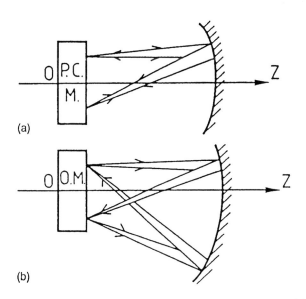

Fig. 7. Optical paths between a concave mirror and (a) a phase-conjugate mirror and (b) an ordinary mirror.

mixture of an ordinary mirror and a phase conjugate mirror. Figure 7 explains intuitively why the two terms on the right-hand side of (30) lead to the same $2d$-value of the period.

More precisely one has $\mathbf{F}(\mathbf{k}, l) = \mathbf{RB}(\mathbf{k}, l)$ by using (24) with the condition $\mathbf{F}(\mathbf{k}, 0) = 0$. It follows that the self-oscillations amplitudes $\delta\epsilon_f$ and $\delta\epsilon_b$ are connected in $z = l$ by a relation of type

$$\delta\epsilon_f = \frac{1}{2}e^{i\phi}\{(R_{11} + R_{22}^*)\delta\epsilon_b + (R_{12} + R_{21}^*)\delta\epsilon_b^* + \delta\epsilon_c\}, \tag{32}$$

where $\delta\epsilon_c$ is a linear combination of components $b(\pm\beta, \pm\mathbf{k})$ and $b^*(\pm\beta, \pm\mathbf{k})$, which can be written in terms of two "extra" fields

$$\delta\epsilon_c = (R_{11} - R_{22}^*)\delta\epsilon_{b,ex} + (R_{21}^* - R_{12})\delta\epsilon_{b,ex}^*. \tag{33}$$

At the threshold, our study proves that the complementary field $\delta\epsilon_c$ has an amplitude much smaller than the two other fields of (32) when T_2 is not too small. For example, for $\Delta = 400$ in Fig. 4, one has

$$\delta\epsilon_F = .5\delta\epsilon_B + .45\delta\epsilon_B^* + .16\delta\epsilon_{B,ex}^*. \tag{34}$$

The small extra field $\delta\epsilon_c$ is responsible for the fact that the threshold period was not exactly equal to $2d$, but was generally 10 percent higher.

One can, therefore, conclude that in this experiment, the sodium cell acts as a mixture of both an ordinary mirror and a phase-conjugate mirror.

2.6 Role of the Homogeneous Dephasing Time

It is essential to describe the temporal aspect of the bifurcation, then to analyze the behavior of the feedback mirror device as a resonator. The role of T_2 has been investigated first, so as to test the adiabatic approximation, which consists of eliminating the polarization P by neglecting its time derivative in the Bloch equations. This approximation is supposed to be valid within the limits of small ratio T_2/T_1. In the Tucson experiment, instabilities have been observed for Argon pressure varying between 0.3 Torr ($T_2 \approx 22$ ns) up to 22 Torr ($T_2 \approx 1$ ns), i.e., for $0.1 < T_2/T_1 < 1.5$. Therefore, one could hope that for the smallest values of T_2, the adiabatic approximation would be valid. Indeed, we have first performed a numerical investigation in this framework. We have calculated the propagation of a cw Gaussian field incoming on the two-level cell at time $t = 0$, going through the cell and returning to it via the mirror, neglecting the longitudinal grating ($|E_f + E_b|^2 = |E_f|^2 + |E_b|^2$) and the diffraction inside the medium, which allows us to treat analytically the propagation inside the medium. The spatio-temporal evolution of the electric field $E(r, t)$ displayed far-field profiles in good agreement with the observations, but temporal aspect was in total disagreement with the observations, because the oscillation period was always higher than $3d$, and varied within a factor of 3, according to the value of the round-trip time d of the optics and other physical parameters; therefore, these results were not published.

The linear stability analysis was used later to investigate the effect of the atomic dephasing time. The analysis showed that the main error in our numerical calculation was to make the adiabatic approximation. It also shows that the adiabatic approximation is valid only for cases in which T_2 is smaller than $0.1T_1$, which was not the case in our experiment. The role of T_2 was investigated in two ways.

The first approach consisted of keeping the *on-resonance absorption length* $\alpha_0 l$ *constant* and decreasing the ratio T_2/T_1. The value of the absorption coefficient $\alpha_0 l$ was chosen from the experimental value in Fig. 3, $\alpha_0 l = 5000$. The threshold curves, or marginal stability curves ($\mathcal{I}_0$, versus θ_D) are drawn in Fig. 8 for the two cases $T_2/T_1 = 1$ (left curve), and 0.01 (right curve). In these two cases, the threshold is $\Delta \approx 400$ and the angular selectivity is good, the difference concerns the threshold periods: For $T_2 = T_1$, the period is clearly well defined, $T \approx 2d$ around the highest maximum (minimum threshold in the instability domain, Fig. 4); while for the small T_2, the period changes by a factor of two in close vicinity to the maximum (for a detuning variation of 10 percent), and is $T = 3d$ at threshold. The marginal stability curves are identical for any $T_2/T_1 < 0.1$. Therefore, first the linear analysis shows that the adiabatic approximation is valid for T_2 smaller than $0.1T_1$, and second, we understand that modeling the sodium gas by a Kerr material was not adapted to the experiment, where T_2 was greater than $0.1T_1$.

In a real experiment, the way to decrease T_2 is to increase the buffer gas pressure, keeping constant the sodium concentration. Let us take the parameters values of Fig. 4 of [Tai, Giu], which displays a $2d$ oscillation: $T_2 = 0.3T_1, N \approx 10^{12}$ atoms/cm^3, or $\alpha_0 l = 240$. The instability domain shown in

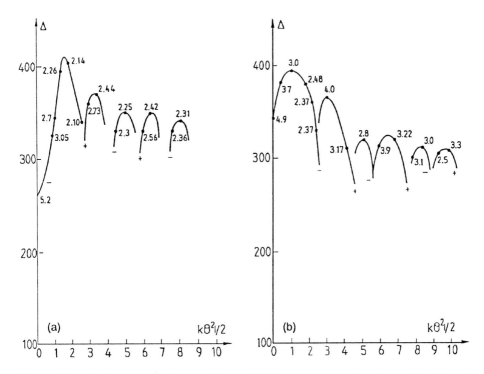

Fig. 8. Threshold detuning for periodic instability versus θ_D, for $\mathcal{I}_0 = 30,000$ and $\alpha_0 l = 5000$, (a) $T_1 = T_2 = 2L/c$, (b) $T_2 = 0.01T_1, T_1 = 2L/c$.

Fig. 9, which corresponds to the latter case, has noticeably shrunk with respect to Fig. 4. Moreover, no instability is predicted for $\Delta > 23$, in agreement with (28). This feature agrees with the observation that a large buffer-gas pressure inhibits the instability process. Finally, in the domain of sodium concentration of $10^{12} - 10^{13}$ atoms/cm³, our study predicts an instability in the feedback mirror device if the buffer gas pressure is lower than about 20 Torrs, as indeed was observed.

In this domain of concentration, while the ratio T_2/T_1 can be as small as 0.1, the observed $2d$ period cannot be reproduced within the adiabatic approximation, as it is when including the term $\frac{1}{T_2}\frac{dP}{dt}$ in ($4b$). In fact, for T_2/T_1 larger than 0.3, the reflectivity matrix elements display the properties $R_{11} \approx R_{22}^*$ and $R_{12} \approx R_{21}^*$, which ensure that the two forward components at frequencies $\pm\beta$ are connected to the backward components by similar relations

$$f(\pm\beta) = R_{11}b(\pm\beta) + R_{12}b^*(\mp\beta), \tag{35}$$

leading to the simple property stated in (31).

The reason for such a result is not clear. In fact, neglecting the role of T_2 seems to prevent any mirror effect. What is more surprising is the revival of the mirror effect due to the two combined processes of the polarization and the population relaxations.

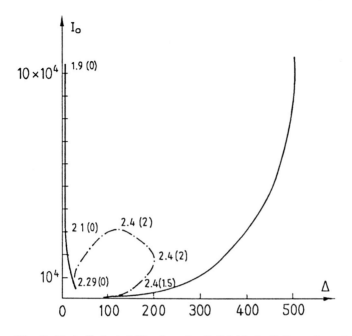

Fig. 9. Periodic instability domain. Solid (dashed) line when the longitudinal grating is included (suppressed) in the treatment.

2.7 Role of the Time Delay

Let us finally mention that the threshold characteristics are unchanged when the mirror is shifted far away from the cell or toward the cell, as it was in the experiment. The threshold period is always found about $2d$, even when the distance is reduced or increased by a factor of 10. Therefore, *the gain seems to be better governed by the optical constraints than by the usual Raman or Rayleigh gain process*, as it was invoked for interpretation of the ring cavity experiment by Khitrova and colleagues [Khi], or the Ségard and Macke experiment [SeM, SML] as will be discussed in the next sections.

3 The Ségard and Macke Experiment

3.1 Experiment

In the Ségard and Macke experiment, a cw laser beam is sent through a waveguide of length L equal to 182 m, with two mirrors at the ends filled with HCN molecules, which behave like homogeneously broadened two-level atoms (Fig. 10). The observed frequency roughly varies between $c/4L$ and $c/2L$, as the mistuning is varied. This experiment was interpreted as a result of a Rabi (also called Raman) gain, the threshold being reached when the generalized Rabi frequency associated with one pass intra-cavity intensity approaches the mistuning between the laser frequency and the adjacent cavity mode (rule of thumb).

The observed instability is then "multimode" [SeM, SML, LuN]. Let us recall that it is customary to distinguish single-mode and multimode instabilities, according to which longitudinal cavity mode becomes unstable, the resonant mode (the mode nearest to ω_L) or another side mode, respectively. A multimode instability can arise only if the power broadened atomic line width is of the order or larger than the free spectral range (here $c/2L$), depending only on which cavity mode lies in the frequency regions characterized by the presence of gain. Our analysis will discuss the point of view of [SML], which interprets the unstable mode in terms of gain via the rule of thumb.

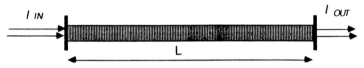

Fig. 10. Fabry-Pérot cavity device.

3.2 Linear Analysis

The experimental situation for the occurrence of this multimode instability appears as *ideal* for a theoretical interpretation. One of the reasons lies in the irrelevance of the Doppler broadening. Another feature that simplifies this study is the aspect of the instability, which is only temporal, because only one transverse mode of the Fabry-Pérot interferometer is excited. Therefore, the linear analysis is performed for on-axis perturbation ($\mathbf{k} = 0$).

The exact linear stability analysis with $\mathbf{k} = 0$ has been performed previously by Van Wonderen [VaW, WSu], taking into account the reflectivity of the mirrors ($R \approx 0.68$ in the experiment) and the absorption of the pump fields. Van Wonderen has obtained much better agreement with the experimental result than Ségard [SeM, SML] with the "equivalent" ring cavity device, showing the importance of the standing-wave effects.

We have checked that a simplified linear stability analysis [MLe], assuming $I_f(z) = I_b(z) = I_0$ and taking into account a reflectivity $R \neq 1$ in the boundary conditions, was in good agreement with Van Wonderen's results. Our simplified treatment has the advantage of enabling us to discuss semi-analytically the threshold characteristics. We have investigated especially the gain properties of the medium for the frequencies close to those of the Hopf bifurcation and the physics of the coupling between the forward and backward self-generated fields to understand why taking into account standing-wave effects leads to better agreement with the experiment.

The Fabry-Pérot conditions are

$$\mathbf{B}(l) = \mathbf{F}(l), \quad \mathbf{F}(0) = R\Theta\mathbf{B}(0), \tag{36}$$

with Θ given by the 2×2 matrix

$$\Theta = \begin{pmatrix} e^{i(\phi+\theta_0)} & 0 \\ 0 & e^{-i(\phi+\theta_0)} \end{pmatrix}, \tag{37}$$

where ϕ is a complex phase equal to the sum of the forward and backward complex phases for the stationary amplitudes ($\phi = \phi_f + \phi_b$, with $E_f(z) = E_0 e^{i\phi_f z}$ and $E_b(z) = E_0 e^{-i\phi_b z}$), R is the mirror reflectivity and θ_0 is the cavity mistuning. The relations (22) together with (35) give rise to the characteristic equation for the Fabry-Pérot device

$$D(\lambda) = \det\left(\mathbf{I} - \mathbf{R} - \frac{1}{\mathbf{I} - R\Theta}\mathbf{G}\right) = 0. \tag{38}$$

3.3 Physics of the Coupling

We chose to study the threshold characteristics in the case reported in Fig. 16 of Ségard [SeM, SML], which have been numerically simulated by Van Wonderen [VaW, WSu].

In Fig. 11, the experimental frequency is shown in a solid line, the theoretical threshold frequency deduced from the exact simulation in a dotted line, and our frequency in dashed line for two different reflectivities ($R = 0.7$ and 0.9). The isolated point corresponds to the ring cavity simulation. It appears that our no pump-depletion model agrees with the exact simulations of Van Wonderen, consequently, we can take out some information from our analytical description.

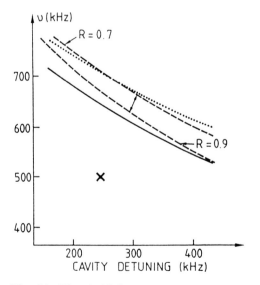

Fig. 11. Threshold frequency versus detuning; comparison between experiment and theory.

Let us investigate the coupling between the small amplitude fields **F** and **B**. The examination of the propagation matrix [LBR, MLe] elements (cf. (38)), for various threshold values, allows us to conclude that the unidirectional coupling $f_-^* \leftrightarrow f_+$ or $b_-^* \leftrightarrow b_+$ does not prevail, as illustrated by the discrepancy between the unidirectional model (cross in Fig. 11 and the correct one). At threshold, the matrix displays a coupling between counter-propagating waves generally stronger than the coupling between copropagating waves [MLe].

It follows that the Fabry-Pérot device cannot be modeled by a ring cavity, even when adjusting the length and the input intensity, because the coupling phenomena is strongly different in the two devices. The simulation of Ségard and colleagues is nevertheless interesting, because it predicts an oscillation at the Rabi frequency in the plane-wave ring cavity device, with copropagating waves. Let us also mention that multiconical emission is predicted with various frequencies near the Rabi range by Patrascu [Pat] in a ring cavity with a quasi-confocal optical arrangement.

3.4 No Rabi Gain at Threshold

The characteristic equation (37) is drastically distinct from the feedback mirror threshold relation (28), which only involves the reflectivity matrix. Equation (37) appears to be generally fulfilled when $\mathbf{R}_{ij} \ll 1$ and $\mathbf{G}_{ij} \approx 1$. These latter relations emphasize the small reflectivity properties of the intrinsic medium and on the finite value of the gain matrix elements (24). These conditions also drastically differ from the one for the intrinsic system [MLe], which requires an infinite gain. Let us recall that near threshold, the intrinsic system gain has a broadband Rabi peak extending between the two extremes of the generalized Rabi frequency for counter-propagating pump waves, δ and

$$\Omega_{R,max} = \sqrt{4\Omega_0^2 + \delta^2}, \tag{39}$$

where Ω_0 is the Rabi frequency for one atom plus one beam of amplitude E_f, and δ is the detuning. Moreover, in this band the reflectivity is smaller than unity. For the Fabry-Pérot device the gain characteristics at threshold are drawn in Fig. 12, which confirms $\mathbf{G}_{ij} \approx 1$ for a frequency close to the Hopf frequency. The reflectivity matrix elements (26) are not drawn, but were found to be much smaller than unity. Therefore, the interpretation of the self-oscillation in terms of the Rabi gain in Fabry-Pérot is not as simple as for the intrinsic system [MLe].

At this stage, one can assert that while self-pulsing in the Fabry-Pérot interferometer occurs at a frequency near $\sqrt{\Omega_0^2 + \delta^2}$, as for a one-way device, it does not result from large convective gain, but because at this frequency the reflectivity matrix elements are small compared to unity, and the gain matrix elements are of the order of unity. The very pleasant rule of thumb that underlines the systematic correlation between the Rabi frequency and the side-mode mistuning is actually not so trivial. *The self-oscillation defined by (37) results from the compromise between the propagation and the boundary conditions.*

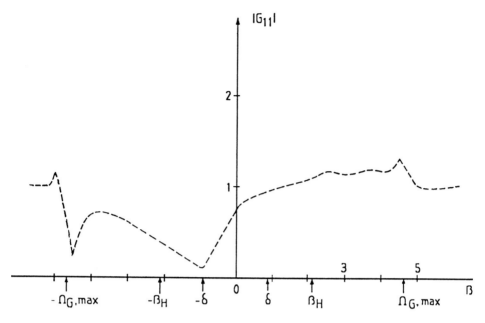

Fig. 12. Threshold gain $|G_{11}|$ in the Fabry-Pérot device, $\mathcal{I}_0 = 1420$, $\Delta = 15$, $\alpha_0 l = 182$, $\theta_D = 0$, $\Psi = -1.6$.

4 Rayleigh Self-Oscillation in an Intrinsic System

The first experiment reporting a temporal instability in a gas driven by counter-propagating pump beams without any cavity (see Fig. 13, called intrinsic system or self-feedback device) was done by Khitrova and colleagues [Khi] on the defocusing side of the D_2 line of sodium vapor without foreign gas. They observed a *Rayleigh* gain lasing (the observed beat frequency was 10.5 MHz, i.e., $\beta T_1 \approx 1$, if T_1 is the radiative lifetime for the D_2 line) without any deformation of the input transverse profile.

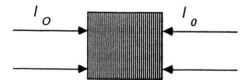

Fig. 13. Intrinsic system diagram.

The linear analysis developed within the framework described above, predicts [LBT] for approximately the same off-resonance absorption coefficient αl, a threshold angular frequency of approximately $65\ T_1^{-1}$, i.e., about two orders of magnitude higher than the observation. In a real situation, several mechanisms may be relevant for defining the period of the instability that were not included in the above linear stability analysis equations. Among those mechanisms, we have selected two effects, the pump depletion [LBT] and the Doppler effects, both of which work to remove the discrepancy between theory and experiment.

Actually the experiment is difficult to interpret, because the observation is reported for a unique laser frequency and intensity, nevertheless, from the experimental details given in [LBT], we assume that the observation is close to the Hopf bifurcation threshold. Moreover, there is no precise indication concerning the atomic concentration and the optics.

The experimental data are summarized now and must be kept in mind to interpret the experiment. The power of the laser was 300 mW, it corresponds to an intensity $\mathcal{I}_0$ scaled to the saturation intensity of about 4×10^4 at each extremity of the cell, if the optics are the same as in Sect. 2, but $\mathcal{I}_0$ may be ten times smaller if the input radius is three times larger. The inverse of the absorption length is $\alpha_0 = 3.3x10^{10}N$, where the concentration N is expressed in cm^{-3}.

The intrinsic system schematized in Fig. 13 has boundary conditions

$$\mathbf{F}(0) = \mathbf{B}(l) = 0. \tag{40}$$

The threshold characteristics within the no pump-depletion model are determined by

$$D(\lambda) = \det(\mathbf{A}_1 + \mathbf{A}_2) = 0, \tag{41}$$

which results from (23) and (39), and also can be considered as the condition for *infinite gain and reflectivity*, because the determinant of the two matrices $\mathbf{G}$ and $\mathbf{B}$ in (24, 25) is proportional to the inverse of expression (41).

Bar-Joseph and Silberberg [BrJ] predicted the occurrence of a Raman self-oscillation in a situation of no dephasing collisions, $T_2 = 2T_1$, and interpreted this self-lasing as a result of the enhancement of the Rabi emission process. Actually, the experiment was set up to observe a Raman lasing.

We first present in Sect. 5.1 the threshold characteristics with the Bar-Joseph and Silberberg model for a large range of detunings $\Delta = 100$ to $\Delta = 525$, keeping $\alpha l = 0.14, l = 10$ cm, and $T_2 = 2T_1$. This shows that the Hopf frequency is not in the Rayleigh domain, but is also clearly outside the strong gain frequency range. The depletion of the pump is introduced in Sect. 3.2. The main effect of the pump depletion is a strong decreasing of the Hopf bifurcation frequency, giving rise to $\beta T_1 \approx 1$ for $\alpha l \approx 1$. The transverse Doppler effect also is shown to decrease the frequency (Sect. 3.3).

4.1 Characteristics of the Self-Oscillation in the No-Pump Depletion Model

Threshold Frequency versus Medium Parameters. The lowest threshold corresponds to a dynamical instability with uniform transverse profile, i.e.,

$|\mathbf{k}| = 0$. Figure 14 displays the dynamical instability domain (threshold intensity versus detuning) for $T_1 l/c = 50$ (cell length $l = 10$ cm) and $\alpha l = 0.14$. It is composed by three distinct branches, but the lowest threshold intensity is surprisingly constant as the detuning varies by a factor of 5. The threshold frequency is printed along the curves. It appears that the Raman instability is obtained for small detuning only ($\Delta < 150$), which can be easily understood, because the atomic nutation is efficient when the detuning is smaller than (or of the order of) the Rabi frequency. This remark explains why the *threshold frequency crucially depends on the detuning*. In consequence, the results predicted in [BrJ] have no universal characteristics for intrinsic gaseous cells without buffer gas.

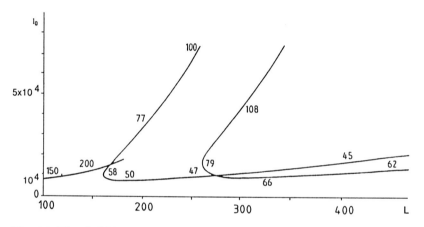

Fig. 14. Threshold input intensity scaled to the on-resonance saturation intensity, versus detuning.

When the longitudinal grating is suppressed in the expression of the A_{ij} matrix elements, the instability domain is reduced, but the results concerning the threshold frequencies are unchanged [LBT], as they were in the case of the feedback mirror device. This property will be used in the following subsection, because analytical calculations are tractable within this hypothesis only.

For the experimental detuning, $\Delta = 525$, Figs. 15a and b show that the predicted frequency is 65 times larger than the experimental frequency. The threshold intensity I_0 is reported in Fig.15a as a function of l/c, and the corresponding frequencies in Fig.15b are labeled with the same letters a,b, ... e. The solid line represents the instability boundary, because outside this domain, the maximum growth rate is negative (the system is stable), while inside this domain the maximum growth rate is positive. The frequency of the upper branch "e" is nearly equal to the maximum of the Rabi frequency, which is drawn in a dashed line for comparison, but the corresponding intensity is too high to be reached by the experiment. The threshold intensity of the lowest branch ($\approx 16 \; times \; 10^{33}$) is independent of the cell length, for any sodium cell length smaller than 25 cm; its frequency (curve "a") is $\beta T_1 \approx 65$, which is four times smaller than the detuning, and clearly much larger than unity. *Therefore, the lowest branch can neither be associated to a Raman gain nor to a Rayleigh gain process.*

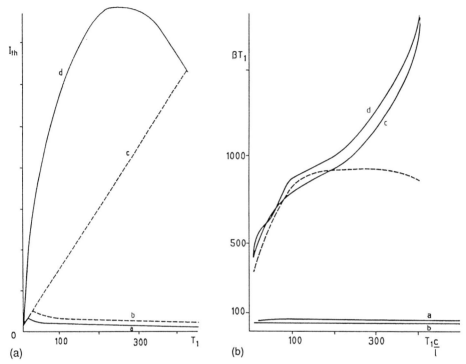

(a)

(b)

Fig. 15. Solutions of (41): input intensity (a) and frequency (b) versus cT_1/l, when the longitudinal grating is neglected. $T_2 = 2T_1, \Delta = 525, \alpha_0 l = 40,000$. The threshold corresponds to the curve (a) in both figures, with frequency $\beta T_1 = 65$.

Hopf Bifurcation and Gain Properties. Finally, let us discuss the dressed-atom emission properties of the intrinsic system. We choose the case of a Raman self-oscillation predicted by Bar-Joseph and Silberberg, where $\Delta = 100, \alpha l = 0.1$, and $T_2 = 2T_1$. Indeed, for this case of pure gas, the atomic nutation effects can be responsible for a Rabi gain.

Self-oscillation occurs in the dashed area in Fig. 16a in the 2-D parameter space $(I_0, l/c)$, where I_0 and l are the input intensity and the cell length, respectively. The threshold frequency (solid line in Fig. 16b) is found to be slightly above the upper limit of the Rabi band (38) drawn in a dashed line, which may lead to the conclusion that the instability results from Rabi gain. Figure 17 displays the gain for a forward input probe field commonly defined as

$$g = 2\ln(f_+(l)/f(0)), \tag{42}$$

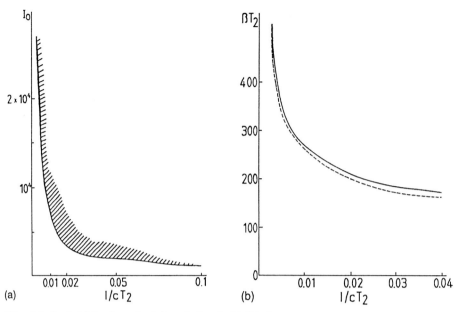

Fig. 16. Instability domain (a) and threshold (b) frequency versus transit time scaled to T_1. $\Delta = 100, \alpha_0 l = 1000, T_2 = 2T_1$.

or $g = 2\ln(G_{11}f_+(0))$, if a probe is sent with frequency β only. This curve clearly does not look like a one-way device curve [BrJ, MLe]. In fact, the solution of the propagation equation (23) with a unique input probe $f(0)$ is of the form

$$f_+(z) = (a_1 e^{s_1 z} + a_1' e^{s_1^* z} + a_2 e^{s_2 z} + a_2' e^{s_2^* z}) f_+(0). \tag{43}$$

Positive gain may arise via two different processes, either if some coefficients a_i take large values, which corresponds to large gain everywhere in the cell, or if $s_{1,2}$ has a positive real part, which corresponds to exponential z-amplification. We have checked that the two processes are present in the case of Fig. 17. The gain is infinite at the Hopf frequency $\beta_H l/c \approx 4.2$, (which is located at a frequency slightly larger than the maximum of the Rabi frequency $\Omega_{R,max} l/c = 3.9$), because $a_i(\beta_H)$ diverges when (40) is fulfilled, while $\text{Re}\{s_{1,2}(\beta_H)\} \approx 0.1$, which means that the probe $f(\beta_H)$ displays negligible exponential gain.

On the contrary, the dashed domain in the whole range of the generalized Rabi frequency $(\delta, \Omega_{R,max})$ corresponds to a strong exponential increase of the probe, with $s_1 \approx 10$. The emergence of strong Rabi gain is found for a large intensity range, well below the Hopf bifurcation. *Therefore, it appears that the occurrence of a periodic temporal instability should not be confused with the "gain" notion as commonly used* [LBT]. Clearly, the linear analysis is not valid in the presence of large exponential gain, so that the above analysis raises questions about what would occur in a real experiment. Two Raman or Rabi lasing processes are possible, either a Rabi lasing built by propagation and initiated by the quantum noise, i.e., the spontaneous emission, or a quasi-Rabi lasing resulting from deterministic behavior of the nonlinear system.

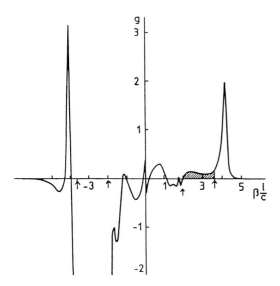

Fig. 17. Probe gain $g = 2\log(|G_{11}|)$ versus frequency scaled to the transit time equal to $0.02T_2$; $I_0 = 3400$, near the threshold of the intrinsic system. The arrows locate the extremes of the generalized Rabi frequency.

4.2 Threshold Characteristics for Depleted Pump Fields

In this section, the longitudinal grating is neglected. The implicit solution for the forward pump intensity is

$$I_F(z) = I_0 \exp\left[-\alpha z + I_0 - I_F(z) + A^2\left(\frac{1}{I_F(z)} - \frac{1}{I_0}\right)\right], \qquad (44)$$

which generalizes the Icsevgi and Lamb single-pass solution [IcL] to counter-propagating beams ($I_0 = |\epsilon_0|^2$). For a large range of the absorption coefficient and of the input intensity, the total pump intensity ($I_F(z) + I_B(z)$) can be considered as a constant. In this paper, the pump depletion will be introduced in the following framework: The total pump intensity is assumed to be independent of the location in the cell and equal to $2I_0$, but the difference ($I_F(z) - I_B(z)$) is allowed to vary from 0 at the cell center to ($I_0 - I_B(0)$) at the cell entrance. For the sake of simplicity, the nonlinear pump-depletion effects are neglected, which implies $\alpha l \ll I_0$, when looking at the general solution (43). The propagation of the two-components' perturbation of field vectors **F** and **B** is generally not easy. Nevertheless, in the case of a threshold frequency much smaller than the detuning, a treatment is proposed in [LBT].

The effect of the pump depletion is displayed in Fig. 18, where the threshold frequency is reported as a function of the concentration. The other parameters are those of the experiment, $l = 1$cm ($T_1 c/l = 480$, $\Delta = 525$, $T_2 = 2T_1$). The main point is the lowering of both the threshold intensity and frequency when the pump depletion is introduced. In the whole domain $\alpha l > 0.1$, the solid line (with pump depletion) is lowered by a factor of about 8 with respect to the dashed line (no pump depletion); the frequency βT_1 approaches unity, the experimental value, for $\alpha l \approx 1$. Even for a small value of the absorption, for example, $\alpha l = 0.14$, the frequency βT_1 reduces from 65 to about 6.5. This result is not surprising. The dispersion relation, which determines the threshold frequency, is indeed very sensitive to any change, even a very small change. Therefore, a difference $(I_F - I_B)$, for instance, of a magnitude of order $10 I_0$, can induce a very strong difference in the instability threshold characteristics.

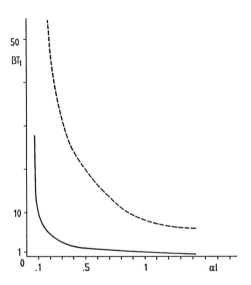

Fig. 18. Threshold frequency versus absorption.

For $\alpha l \approx 1$, where the threshold frequency approaches the experimental value, the approximation $(I_F + I_B) = \text{const} = 2 I_0 e^{-\alpha l/2}$ is still satisfied: The sum $(I_F + I_B)$ varies, in fact, between 5200 at the cell entrance and 4600 at the center, which makes a variation of about 10 percent, which is still reasonable, making our assumption still available. The difference $(I_F - I_B)$ varies between 2400 at the cell entrance and 0 at the center. This large variation explains why the introduction of the pump depletion induces strong modifications of the threshold characteristics.

Drastic effects related to a nonvanishing difference $(I_F - I_B)$ were already reported for phase conjugation experiments in sodium vapor [GKP, KTP].

In conclusion, the main effect of the pump depletion is to strongly lower the value of the threshold frequency by a factor of magnitude of order ten: A threshold angular frequency approximately equal to $(T_1)^{-1}$ occurs for a concentration such that the off-resonance absorption coefficient, αl, is about unity, i.e., for concentrations much larger than those used in the single-feedback mirror experiment [Tai, Giu, LBR]. Even if there is often a large uncertainty on the measure of the concentration, it appears difficult to presently conclude that the pump depletion is the only effect responsible for the onset of the Rayleigh instability observed by Khitrova [Khi].

4.3 Doppler Effect

The atomic motion might not be negligible in the experimental conditions [Khi], because the Doppler line width $(T_2^*)^{-1}$ is of a magnitude of order 2 GHz (the detuning is 2.6 GHz).

One of the effects of the atomic motion may be to wash out the longitudinal grating [LBT] created by the parallel counter-propagating pumps. It arises when the mean atomic velocity v_z is fast enough for $k v_z$ to be much larger than the nutation angular frequency ($\Omega T_2^* \ll 1$). Then, integrating over atomic velocities is straightforward when neglecting the longitudinal grating. Such *Doppler broadening* was investigated in the case of one-way propagation. The main conclusion was that it drastically enlarges the Rabi sidebands, but only slightly affects the Rayleigh gain [Khi]. Nevertheless, it might be different for atoms submitted to counter-propagating waves. We have recently introduced Doppler broadening in our linear analysis. The threshold frequency was found to decrease by a factor of 2 when the ratio T_2^*/T_2 decreases to 0.1. The experimental ratio is of the order 2×10^{-3}, but we did not pursue our investigation, because the linear analysis became inextricable. Nevertheless, we have clearly observed that the Doppler effect is inclined to reduce the threshold frequency.

Finally the combination of the two effects, pump depletion and Doppler effect, was investigated. While each of them separately reduces the threshold frequency, it seems that taking into account the two effects does not lead to a further reduction of the threshold frequency. Finally, while the above results are in favor of a better agreement between the theory and the experiment, it is not obvious that the combination of the pump depletion and the Doppler effect is actually the key to the discrepancy previously found between the Rayleigh self-oscillation observed in the experiment and the much higher value ($\beta T_1 \approx 65$) predicted by the simplified model. In fact, it would be desirable to get complementary experimental results concerning the variation of the threshold characteristics, input intensity, and frequency as a function of the concentration and to compare them with the theoretical predictions.

5 Conclusion

The study of these three examples allows us to reach some conclusions concerning the method we have used, and also the physical mechanisms involved in the experiments that we have described.

(a) The main conclusions concerning the method are the following: When there is some observation, comparison with any model is much easier if the experiment is described for a large range of parameters. In practicality, the theoretician needs "global" experiments to, first, identify the nature of the bifurcation and, second, to set up a suitable model.

The linear analysis appears as a necessary preliminary stage, which allows us to set up the correct model. For the description of spatio-temporal instabilities, simple models without any experimental support must be used cautiously. For example, the Kerr model, which is very attractive for its simplicity, is unable to bring out the physics in the three cases treated here. Indeed, the polarization relaxation induces the feedback mirror device to behave as a resonator, the Fabry-Pérot device modelized by a Kerr medium would certainly not provide the Rabi oscillation, and the counterpropagation in Kerr media was proved recently to exhibit self-oscillation with a period of the order of a few transit times [Cha], which is very different from a Rayleigh oscillation.

However, the predictions of the linear analysis are limited. It is a first order perturbative treatment valid only in the vicinity of the instability threshold. If the temporal aspect of a Hopf bifurcation is satisfactorily described, on the contrary, spatial patterns are roughly predicted. In the feedback mirror experiment, one needs to remove the angular degeneracy by doing higher order perturbation treatment of the ring patterns predicted by the linear analysis. This can be done by deriving amplitude equations [Man] and should lead to the predictions of one, two, three, or more spots distributed on the far-field ring pattern predicted at order one [PeP, Fir, Ale, Cha, Ged, FiP, GrP, VF, MAI]. The derivation of amplitude equations for Kerr media [PeP, Tai, Giu, AVI, Vor, MMN, MAM, Fir, Ale, Cha, Ged, FiP, GrP, VF, MAI] indeed lead to hexagonal structures. Unfortunately, in the present case of gaseous media, where three-dimensional propagation is involved, analogous treatment appears difficult. As for numerical investigations, they are already very heavy in Kerr systems for three-dimensional propagation [MAM, Fir, Ale, Cha] and presently seem impossible for gaseous media, with the additional difficulty of a delayed feedback in cavity devices.

(b) Let us return to understanding the physics, especially to the problem of the interpretation of the self-generated frequencies. We have questioned the intuitive idea that such optical devices, including nonlinear medium lighted by pumps are susceptible to emit in the frequency range where the nonlinear medium displays large positive gain. As a general rule, it was shown that the self-oscillation is not due to the presence of large gain in cavities. This result may suffer some counterexamples, but it was stated and checked for the ring cavity, the Fabry-Pérot, and the single-mirror devices. On the contrary, the dispersion relation clearly indicates that the gain matrix elements $\mathbf{G}_{ij}$ are of the order of unity at threshold in the Fabry-Pérot (37) and in the ring cavities [Pat]. The threshold gain curve $g(\beta) = |G_{11}(\beta)|$ in the ring cavity is so flat that the medium could be considered transparent for any incident probe field. Then, if the process of amplification by propagation is eliminated, what is the cause of the emission of some frequencies rather then others? While we have no general explanation, we have recently proved [Pat] that constructive interference exists in pure atomic media. This property was stated in the ring cavity where the light propagates forward only: The propagation matrix elements, such as h, t in (22), were shown to display a simple phase-matching condition, ensuring the maximum coupling efficiency between the two components f_+ and f_-^*. Finally, the instability seems more likely due to phase effects than to amplitude effects related to high gain.

For the half-cavity device, the dispersion relation (27) does not involve the gain matrix, but points out that the medium has to get reflectivity matrix elements $\mathbf{R}_{ij}$ of order unity. In fact, the original interpretation [Tai, Giu] of the half-cavity device as a phase-conjugate resonator was not so erroneous, but it was a simplification of the real process. From optical arguments, it is easy to understand the role of the medium, as a mixture of ordinary mirror and phase-conjugate mirror, because both lead to the $c/4l$ self-oscillation. While intuitively the condition $|\mathbf{R}_{ij}| \approx 1$ might explain the mirror-like behavior of the nonlinear medium, it is not enough. One must add some phase relations between the $\mathbf{R}$-matrix elements in order for the two forward components f_+ and f_- to reflect the two corresponding backward components in a similar way $R_{11} = R_{22}^*$; $R_{12} = R_{21}^*$. We have shown that these additional relations can be obtained only if the polarization dephasing time is of the order of T_1. Finally, the mirror-like behavior of the medium arises from the interplay of a simple boundary condition $|\mathbf{R}_{ij}| \approx 1$, and some phase relations between the $\mathbf{R}$-matrix elements, which ensure the phase-matching between the two $\pm\beta$-spectral components of the self-oscillation.

Why do these specific phase-matching relations occur especially when the buffer gas pressure is low? They were not found for T_2 much smaller than T_1, in the ring cavity or in the half-cavity device. This property may be partially understood with the help of the following remark. In our studies, the population decay time T_1 was of the order of magnitude of the cavity round-trip time. In the adiabatic approximation, the atomic response, which is proportional to the polarization or else to $(1 + i\Delta)\epsilon w$, is dephased with respect to the electric field, without any hope to get matching, because T_1 acts alone. Phase-matching effects can only arise via another dephasing process, introducing a new degree of liberty in the system. The role of the polarization dephasing time T_2 is, in fact, to reset some phase in the field components. The role of T_2 may be neglected for the description of purely spatial instability [Fir, Ale, Cha, Ged, FiP, GrP, VF, MAI], but this study proves that it plays an important role in the onset of spatio-temporal instability, this explains why the Kerr model is not always suitable for two-atom systems.

The origin of the instabilities in the case of the *intrinsic device* appears more ticklish. This system without mirror would be a priori the best candidate to display instability resulting from high gain, because the dispersion relation is equivalent to infinite gain g^i. The linear stability analysis shows that this system without mirror is quite intricate. It may display self-oscillation with frequency close to the Rayleigh and Rabi bands characterized by large positive gain. But first, the quasi "Rabi instability" occurs outside the domain characterized by strong z-amplification (exponential enhancement along the propagation axis), while the infinite gain results from (40). Second, the Rayleigh instability was not easily reproduced. The model used previously for analyzing the instability in cavities, curiously, failed for the intrinsic system, predicting a frequency much larger than the observation by about two orders of magnitude. Therefore, we have searched among the imperfections of our model, for some factor that may be relevant. We have checked the small *pump-depletion effects* and the *Doppler effect* and found them to have very strong influence on the Hopf frequency. The role of the pump depletion can be understood when looking at the dispersion relation, which seems very sensitive to small differences between the forward and backward pump intensities.

The instability threshold for the data of Fig. 15 was found to be very sensitive to the transverse direction, i. e., to the transverse grating created by the interference between the pump and the probe. Indeed, the marginal stability curve has a very sharp minimum in the direction of on-axis rays ($|\mathbf{k}| = 0$). This property enforces the intuitive and common idea that the intrinsic system self-oscillates via its own internal feedback. Finally, even if some results remain unclear, such as the strong dependence value of the emitted frequencies upon small effects, the main mechanism of the self-oscillation in the intrinsic system seems quite well understood, because *self-feedback effects are created by the transverse grating.*

Does this sensitivity to small pump-depletion effects question the validity of the results concerning the cavity devices? As a matter of fact, while we have not checked the role of the pump absorption on the threshold period, we can conjecture that small additional terms, such as $I_F - I_B$, may play a secondary role in defining the self-oscillation inside a cavity, where the prime part is taken by the boundary conditions.

References

[Cas]   Casperson, L.W.: Gas Laser Instabilities and Their Interpretation. In Instabilities and Chaos in Quantum Optics II, ed. Abraham, N.B., Arecchi, F.T., Lugiato, L.A., NATO Series B, **177** Plenum Press, N.Y., 1987, 83.

[WAb]   Weiss, C.O., and Abraham, N.B.: Characterizing Chaotic Attractors Underlying Single Mode Laser Emission by Quantitative Laser Field Phase Measurement. In Measures of Complexity and Chaos, ed. Abraham, N.B., Albano, A.M., Passamento, A., and Rapp, P.E., NATO Series B, **208** Plenum Press, N.Y., 1989, 269.

[Are]   Arecchi, F.T.: Shilnikov Chaos in Lasers in Op. Cit. [Cas] 27; and Arecchi, F.T., Lapucci, A., and Meucci, R.: Shilnikov Chaos: How to Characterize Homoclinic and Heteroclinic Behaviour. In Op. Cit., [WAb] 281.

[HLB]   Hennequin, D., Lefranc, M., Bekkali, A., Dangoisse, D., and Glorieux, P.: Characterization of Shilnikov Chaos in a CO_2 Laser Containing a Saturable Absorber. In Op. Cit. [WAb] 299.

[Ike]   For a documentation on recent studies on spatio-temporal effects in nonlinear optics, see the special review in J. Opt. Soc. Am. B **6** June-July, 1990; particularly the paper, Overview of Transverse Effects in Nonlinear Optical Systems. Eds., Abraham, N.B., and Firth, W.J., 948.

[Ida]   Ikeda, K.: Multiple-Valued Stationary State and Its Instability of the Transmitted Light by a Ring Cavity System. Opt. Comm. **30** 1979, 257.

[IkD]   Ikeda, K., Daido, H., and Akimoto, O.: Optical Turbulence: Chaotic Behavior of Transmitted Light from a Ring Cavity. Phys. Rev. Lett. **45**, 1980, 709.

[Gry]   Grynberg, G.: Mirrorless Four-Wave Mixing Oscillation in Atomic Vapors. Opt. Comm. **66** 1988, 321.

[GBV]   Grynberg, G., Le Bihan, E., Verkerk, P., Simoneau, P., Leite, J.R.R., Bloch, D., Le Boiteux, S., and Ducloy, M.: Observation of Instabilities Due to Mirrorless Four-Wave Mixing Oscillations in Sodium. Opt. Comm. **67** 1988, 363.

[PHe]   Pender, J., and Hesselink, L.: Degenerate Conical Emissions in Tomic Sodium Vapor. JOSA B **7** 1990, 1361.

[PeP]   Petrossian, A., Pinard, M., and Maître, A., Courtois, J. Y., and Grynberg, G.: Transverse Pattern Formation for Counterpropagating Laser Beams in Rubidium Vapor. Europhysics Letters **18** 1992, 689.

[KMG]   The observation of one (resp. three) cone centered on the bissector of two (resp. three) pump laser beams is reported by Kauranen, M., Maki, J. J., Gaeta, A.L., and Boyd, R. W.: Two-Beam Excited Conical Emission. 0pt. Lett. **16** 1991, 943.

[SeM]   Ségard, B., and Macke, B: Self-Pulsing in Intrinsic Optical Bistability with Two-Level Molecules. Phys. Rev. Lett. **60** 1988, 412.

[SML]   Ségard, B., Macke, B., Lugiato, L.A., Prati, F., and Brambilla, M.: Multimode Instability in Optical Bistability. Phys. Rev. A **39** 1989, 703.

[Oro]   Orozco, L.A., Kimble, H. J., Rosenberger, A. T., Lugiato, L. A., Asquini, M. L., Brambilla, M., Narducci, L. M.: Single-Mode Instability in Optical Cavity. Phys. Rev. A **39** 1988, 1235.

[Gra]   Granclément, D., Grynberg, G., and Pinard, M.: Observation of Continuous-Wave Self-Oscillation Due to Pressure-Induced Two Wave Mixing in Sodium. Phys. Rev. Lett. **59** 1987, 40.

[Khi]   Khitrova, G., Valley, J. F., and Gibbs, H.M.: Gain-Feedback Approach to Optical Instabilities in Sodium Vapor. Phys. Rev. Lett. **60** 1988, 1126.

[Tai]   Tai, K.: Nonlinear Optical Transverse Effects: CW On-Resonance Enhancement, CW Off-Resonance Interference Rings, Crosstalk, Intracavity Phase Switching, Self-Defocusing in GaAs Bistable Etalon, Self-Focusing and Self-Defocusing Optical Bistability, and Instabilities. Ph.D. Dissertation (University of Arizona, Tucson, Ariz., 1984)

[Giu]   Giusfredi, G., Valley, J. F., Pon, R., Khitrova, G., and Gibbs, H. M.: Optical Instabilities in Sodium Vapor. JOSA B **5** 1988, 1181.

[AVI]   Akhmanov, S. A., Vorontsov, M. A., and Ivanov, V. Yu.: Large-Scale Transverse Nonlinear Interactions in Laser Beams; New Types of Nonlinear Waves; Onset of "Optical Turbulence." JETP Lett. **47** 1988, 707.

[Vor]   Vorontsov, M. A.: The Problem of Large Neurodynamics System Modelling: Optical Synergetics and Neural Networks. In Proc. SPIE **1402** 1991, 116.

[MMN]   McLaughlin, D. W., Moloney, J. V., and Newell, A. C.: New Class of Instabilities in Passive Optical Cavities. Phys. Rev. Lett. **55** 1985, 168.

[MAM]   Moloney, J. V., Adachihara, H., McLaughlin, D. W., and Newell, A. C. In Chaos, Noise and Fractals. Eds. Pike, E.R., and Lugiato, L.A., London, Hilger 1987, 137.

[Fir]   Firth, W.: Spatial Instabilities in a Kerr Medium with Single Feedback Mirror. J. Mod. Opt. **37** 1990, 157.

[Ale]   D'Alessandro, G., and Firth, W.: Hexagonal Spatial Patterns for a Kerr Slice with Feedback Mirror. Phys. Rev. Lett. **66** 1991, 2597.

[Cha]   Chang, R., Firth, J. W., Indik, R., Moloney, J. V., and Wright, E. M.: Three-Dimensional Simulations of Degenerate Counterpropagating Beam Instabilities in a Nonlinear Medium. Opt. Comm. **88** 1992, 167.

[Ged]   Geddes, J. B., Indik, R., Moloney, J. V., Firth, W. J., and McDonald, G. S.: Hexagons and Their Dynamics and Defects in Nonlinear Counterpropagation in Kerr Media. Proc. 18th I.Q.E. Conf., Vienna, 1992.

[FiP]   Firth, W. J., and Paré, C.: Transverse Modulational Instabilities. Opt. Lett. **13** 1988, 1096.

[GrP]   Grynberg, G., and Paye, J.: Spatial Instability for a Standing Wave in a Nonlinear Medium. Europhys. Lett. **8** 1989, 29.

[VF]   Vorontsov, M.A., Firth, W.J.: Pattern Formation and Competition in Nonlinear Optical Systems with Two-Dimensional Feedback. Phys. Rev. A **49** April 1994, 2891.

[MAI]   Moloney, J. V., Adachihara, H. A., Indik, R., Lizarraga, C., Northcutt, R., McLaughlin, D. W., and Newell, A. C.: Modulational-Induced Optical Pattern Formation in a Passive Optical-Feedback System. J. Opt. Soc. Am. B **7** 1990, 1039.

[LeB]   A presentation of various theoretical treatments is given by Le Berre, M., Ressayre, E., and Tallet, A.: Instabilities in Optical Cavities. Inst. Phys. Conf. Ser.115, Sect. 5, ECOOSA, Rome, 1990, 269.

[LOT]   Lugiato, L. A., Oppo, G. L., Tredicce, J. R., Narducci, L. M., and Pernigo, M.A.: Instabilities and Spatial Complexity in a Laser. JOSA B **7** 1990. 1019, and references herein.

[Mol]   Mollow, B. R.: Stimulated Emission and Absorption Near Resonance for Driven Systems. Phys. Rev. A **5** 1972, 2217.

[Har]   Haroche, S., and Hartman, F.: Theory of Saturated-Absorption Line Shapes. Phys. Rev. A **6** 1972, 1280.

[Pat]   Patrascu, A. S., Nath, C., Le Berre, M., Ressayre, E., and Tallet, A.: Multi-Conical Instability in the Passive Ring Cavity: Linear Analysis. Optics Comm. **91** 1992, 433.

[RAF]   Optical Phase Conjugation. Ed. Fisher, R.A., Ac. Press, N.Y., 1983, Chap. 1.8; or Slusher, R.E., in Prog. in Optics. Ed. Wolf, E., North Holland, Amsterdam, **12** 1973, 1.

[Pap]   Papoulis, A.: In Probability, Random Variables and Stochastic Processes. McGraw Hill, N. Y., 1965.

[McC]   McCall, S. L.: Instabilities in CW Light Propagation in Absorbing Medium. Phys. Rev. A **9** 1974, 1515.

[BRZ]   Le Berre, M., Ressayre, E., Tallet, A., and Zondy, J. J.: Linear Analysis of Single-Feedback-Mirror Spatio-Temporal Instabilities. IEEE J. Q.E. **26** 1990, 589.

[BrJ]   Bar-Joseph, I., and Silberberg, Y.: Instability of Counterpropagating Beams in a Two-Level-Atom Medium. Phys. Rev. A **36** 1987, 1731.

[LBR]   Le Berre, M., Ressayre, E., Tallet, A.: Self-Oscillations of the Mirrorlike Sodium Vapor Driven by Counterpropagating Light Beams. Phys. Rev A **43** 1991, 6345.

[Ber] Le Berre, M., Ressayre, E., and Tallet, A.: Gain and Reflectivity Characteristics of Self-Oscillations in Self-Feedback and Delayed Feedback Devices. Opt. Comm. **87** 1992, 358.

[YaP] Yariv, A., and Pepper, D. M.: Amplified Reflection, Phase Conjugation and Oscillation in Degenerate Four-Wave Mixing. Opt. Lett. **1** 1977, 16.

[LuN] Lugiato, L. A., and Narducci, L. M.: Multimode Laser with an Injected Signal: Steady-State and Linear Stability Analysis. Phys. Rev. A **32** 1985, 1576.

[VaW] Van Wonderen, A. J.: Instabilities in Optical Bistability. Thesis, 1990.

[WSu] Van Wonderen, A. J., and Suttorp, L. G.: Instabilities of Absorptive Optical Bistability in a Nonideal Fabry-Pérot Cavity. Phys. Rev. A **40** 1989, 7104; Phase Instabilities in Absorptive Optical Bistability. Opt. Comm. **73** 1989, 165; and Uniform-Field Theory of Phase Instabilities in Absorptive Optical Bistability. Physica A **161** 1989, 447.

[MLe] Le Berre, M., Ressayre, E., Tallet, A.: Physics in Counterpropagating Light Beam Devices; Phase Conjugation and Gain Concepts in Multi-Wave Mixing. Phys. Rev. A **44** 1991, 5958.

[LBT] Le Berre, M., Ressayre, E., Tallet, A.: An Interpretation for Rayleigh Self-Oscillations in a Pure Sodium Vapor Driven by Counterpropagating Light Beams. Phys. Rev. **46** 1992, 4123.

[IcL] Icsevgi, A., and Lamb, W. E.: Propagation of Light Pulses in a Laser Amplifier. Phys. Rev. **185** 1969, 517.

[GKP] Grynberg, G., Kleimann, B., Pinard, M., and Verkek, P.: Amplified Reflection in Degenerate Four-Wave Mixing: A More Accurate Theory. Optics Lett. **8** 1983, 614.

[KTP] Kleimann, B., Trehin, F., Pinard, M., and Grynberg, G.: Degenerate Four-Wave Mixing in Sodium Vapor in the Rabi Regime. JOSA B **2** 1985, 704.

[Man] Manneville, P.: Structures, Dissipatives, Chaos, and Turbulence. Ed. Saclay, A., 1991; or Dissipative Structures and Weak Turbulence. Academic Press, 1990.

5 Transverse Traveling-Wave Patterns and Instabilities in Lasers

Q. Feng, R. Indik, J. Lega, J.V. Moloney, A.C. Newell, and M. Staley

Arizona Center for Mathematical Sciences, University of Arizona, Tucson, AZ 85721, USA

Introduction

Complex pattern formation or destruction is commonly observed in spatially extended, continuous, dissipative systems, such as convection in ordinary and binary mixture fluids and in liquid crystals, Taylor-Coullet flow, chemical reaction, and directional solidification, etc. [NPL]. Rayleigh-Benard convection has been a canonical system in developing and testing ideas for understanding patterns and transitions [NPL]. Lasers provide another physically and mathematically convenient system for studying space-time complexity, such as pattern selection and spatio-temporal chaos, widely observed in hydrodynamic systems. In this article, we will illustrate, using the two-level and Raman single, longitudinal-mode lasers as prototype systems, the rich variety of pattern-forming instability mechanisms that may appear in wide-aperture lasers.

Spatial patterns in the transverse section of lasers have attracted much attention recently [AbF]. A novel discovery, e.g., is the optical vortex state [CGR], which bears an analogy with the spiral waves often observed in chemical reactions. While most theoretical studies have concentrated on lasers with transverse waveguiding confinement or externally imposed transverse spatial modes [AbF], we shall show instead that under relaxed transverse constraints, namely, assuming idealized infinite extension in the transverse directions, lasers can operate in a natural state, taking the form of *transverse traveling waves*, which are exact solutions to the full nonlinear laser equations [Jk1, NwM]. One can imagine that in a large aspect ratio system, for which the pattern wavelength is much smaller than the transverse dimension of the laser cavity, local patches or domains of wave patterns appear from noise as the laser is turned on. It is natural to ask whether this traveling-wave solution is stable or unstable to arbitrary disturbances. The availability of an exact finite amplitude solution to the nonlinear equations describing the physical system of lasers is a rare luxury, and the simplicity of the form of the traveling-wave solution to both two-level and Raman problems allows us to extend the analysis of the laser well beyond threshold. We identify various phase and amplitude instabilities [NPL] and draw stability diagrams in the pumping parameter – wave number space. The stability region in the stress parameter – wave-number space – is known as the Busse balloon in the context of fluid convection [Bus].

Phase instabilities involve the growth of sideband modes with local wavevectors and frequencies in the neighborhood of the traveling wave. Phase instabilities are easily seen by constructing the Cross-Newell phase-diffusion equation [CrN], which has been established for Rayleigh-Benard convection and checked against the Busse balloon [NPS]. For transverse traveling-wave patterns in lasers, the Cross-Newell phase equation was derived recently, and various phase instabilities have been identified [Leg].

Amplitude instabilities are more difficult to identify, because they involve the growth of modes whose wavevectors and frequencies are significantly different from those of the traveling wave. Notable amplitude instabilities are cross-roll instabilities in Rayleigh-Benard convection [Bus] and "short wavelength Eckhaus instabilities" in Taylor-Coullet flow, which originate from interaction among modes with resonant wave numbers [PpR].

In laser systems, we have identified a novel amplitude instability that severely restricts the stability domain (in the pumping parameter – wave-number space) of the transverse traveling waves [FMN]. This amplitude instability is traced back to the interaction between oppositely traveling waves and leads to a transition from a wave traveling in one direction to a chaotic modulation wave traveling in the opposite direction, with an interesting transient. Coexistence of a stable band of traveling-wave solutions, Eckhaus, zig-zag, or Benjamin-Feir phase instabilities and amplitude instabilities at a fixed value of the stress parameter just above the threshold for lasing suggests that extremely rich pattern-forming scenarios are possible.

This survey is organized as follows. We begin in Sect. 1 by introducing the laser models, including transverse degrees of freedom, and reviewing some basic bifurcation behavior of the spatially homogeneous solutions. In Sect. 2, we conduct linear stability analysis of the transverse traveling-wave solutions directly from the fundamental equations and from the Cross-Newell phase equation. Using both models, we compare the stability boundaries in the pumping parameter – wave-number space. In Sect. 3, we present results of numerical simulation, showing that complex pattern evolution is possible just beyond the onset of lasing for both two-level and Raman lasers. In particular, we provide a specific illustration of the weakly turbulent behavior for a 1-D two-level laser and for a 2-D Raman laser. The complex spatio-temporal behavior can be related to the different phase and amplitude instabilities acting in concert. Section 4 contains some concluding remarks.

1 Basic Equations and Transverse Traveling-Wave Solution

The mathematical description of both two-level and Raman lasers derives directly from the Maxwell equations for the optical fields and the appropriate material density matrix equations. The Maxwell-Bloch equations can be written in the dimensionless form [NwM, Jk2]

$$\partial_t e - ia\nabla^2 e + \sigma e - \sigma p = i\delta_1 en,$$
$$\partial_t p - re + (1 + i\Omega)p = -en + i\delta_2 |e|^2 p, \tag{1}$$
$$\partial_t n + bn = Re(ep^*),$$

with detuning $\Omega = (\omega_{atom} - \omega_{cavity})/\gamma_{12} + r\delta_3 \equiv -\Delta_s + r\delta_3$ (γ_{12} = polarization decay rate). In these equations e, p, and n are the appropriately scaled envelopes of the electric field, polarization, and inversion population, respectively (e and p are complex and n is real. The scalings for two-level lasers and for Raman lasers are different). The parameters σ and b are the electric field and inversion decay rates, respectively, and are normalized to the polarization decay rate. In this case, a is the diffraction coefficient, and the transverse Laplacian operator $\nabla^2 = \partial_x^2 + \partial_y^2$ includes the transverse degrees of freedom allowing for pattern formation. The parameter r is proportional to the pumping and acts as the main external control parameter analogous to the Rayleigh number in Rayleigh-Benard convection [Bus]. The essential difference between the two lasers lies in the method of pumping employed to achieve population inversion. A two-level laser inversion for lasing is created via incoherent pumping, whereas in a Raman laser a classic three-wave interaction, involving two optical waves and one material wave, introduces a strong coherence between the pump wave and the laser emission field (amplitude e). For the Raman laser, $\delta_1 = \frac{\delta_1'}{\delta r}$, $\delta_2 = \frac{\delta_2'}{\delta r}$ and $\delta_3 = \delta_3'\delta$ (δ_i', $i = 1, 2, 3$ are fixed parameters), here δ is the detuning of the pump laser from the dipole-coupled off-resonant intermediate state, and it can be tuned either positive or negative. For the two-level laser δ_i, $i = 1, 2, 3$ are zero. (For details of the scalings, see [NwM, Jk2].)

When r is increased, the nonlasing solution $e = p = n = 0$ loses stability and gives way to the lasing state, which takes the form of traveling waves in the transverse direction [Jk1, NwM]

$$e_{tw} = \bar{e}e^{i(\mathbf{k}\cdot\mathbf{x}+\omega\mathbf{t})}, \quad p_{tw} = \bar{p}e^{i(\mathbf{k}\cdot\mathbf{x}+\omega\mathbf{t})}, \quad n_{tw} = \bar{n}. \tag{2}$$

For the two-level laser, $\bar{e}$, $\bar{p}$, $\bar{n}$, and ω are given by

$$|\bar{e}|^2 = b[r - r_0(k)], \quad \bar{n} = r - r_0(k),$$
$$\omega = -\frac{ak^2 + \sigma\Omega}{1+\sigma}, \quad \bar{p} = \frac{r_0(k)}{1+i(\Omega+\omega)}\bar{e}, \tag{3}$$

with the *neutral curve*, at which $\bar{e} = \bar{p} = \bar{n} = 0$,

$$r_0(k) = 1 + \left(\frac{\Omega - ak^2}{1+\sigma}\right)^2. \tag{4}$$

For given wave number k, the traveling-wave solution exists if $r > r_0(k)$, thus the bifurcation is always supercritical. Minimizing $r_0(k)$ with respect to k gives the lasing threshold $r_c = r_0(k_c)$ with critical wave number k_c and critical frequency

$\omega_c = \omega(k_c)$. For the case of positive detuning $\Omega > 0$, which will be considered throughout this article, we have

$$k_c = \pm\sqrt{\Omega/a}, \quad r_c = 1, \quad \omega_c = \Omega, \tag{5}$$

so that the frequency of the electric field is $\omega_{cavity} - \omega_c = \omega_{atom}$, the atomic line frequency. Therefore, by choosing the appropriate transverse structure, the laser can operate at the frequency natural to the medium and the constraints of the end mirrors are removed. The constraints of the end mirrors are replaced, of course, by the constraints of transverse boundaries. Here, however, to concentrate on the main new results, we will assume that by virtue of an appropriate annular (lasing occurs in an annular region between two cylinders) and, therefore, periodic geometry, the constraints of transverse boundaries do not play an important role[1].

For the Raman laser, the bifurcation to the traveling-wave solution (2) can be super- or subcritical, depending on the parameters. The electric field amplitude $\bar{e}$ (real) is now given as the solution to the following quadratic equation in $\bar{e}^2[Jk2]$:

$$\bar{e}^4 + H(r, k^2)\bar{e}^2 + G(r, k^2) = 0, \tag{6}$$

where

$$H(r, k^2) = \left(\tfrac{\delta_1}{b} - \delta_2\right)^{-2}\left[2(\Omega - ak^2)\left(\tfrac{\delta_1}{b} - \delta_2\right) + \tfrac{(1+\sigma)^2}{b}\right],$$
$$G(r, k^2) = \left(\tfrac{\delta_1}{b} - \delta_2\right)^{-2}\left[(\Omega - ak^2)^2 - (r - 1)(1 + \sigma)^2\right].$$

And, ω, $\bar{n}$, and $\bar{p}$ are given by

$$\omega = \tfrac{1}{1+\sigma}\left(\sigma\delta_2\bar{e}^2 - \sigma\Omega - ak^2 + \delta_1\bar{n}\right), \quad \bar{n} = \tfrac{\bar{e}^2}{b},$$
$$\bar{p} = \bar{e}(1 + i\alpha), \quad \alpha = \tfrac{1}{\sigma}(\omega + ak^2 - \delta_1\bar{n}). \tag{7}$$

When $G(r, k^2) = 0$, we have a solution $\bar{e} = 0$ to (6), so $G(r, k^2) = 0$ gives the neutral curve $r = r_0(k)$. For $r > r_0(k)$ (given k), $G(r, k^2) < 0$, and (6) has only one solution

$$\bar{e} = -\frac{H}{2} + \frac{1}{2}\sqrt{H^2 - 4G}. \tag{8}$$

We shall concentrate on the case in which $\delta_3 - \Delta_s > 0$, so that the lasing threshold is $r_c = 1$, $k_c = \sqrt{(\delta_3 - \Delta_s)/a}$. Solution (8) may also exist in the regime in the (k, r) plane, where $r < r_0(k)$, $H^2 - 4G > 0$, and $H < 0$, therefore, we have subcritical bifurcation (another solution $\bar{e} = (-H - \sqrt{H^2 - 4G})/2$ is always unstable due to the subcritical bifurcation). We will show in the next section that the subcritical bifurcation indeed can happen for some parameter sets, which indicates that the Raman laser may exhibit richer pattern-forming scenarios than the two-level laser.

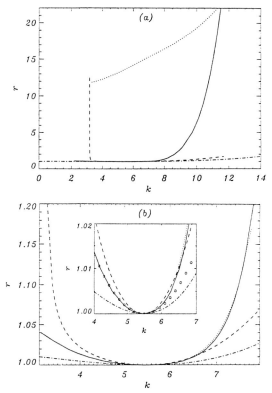

Fig. 1. Stability diagram for a two-level laser. The low-r part of (a) is enlarged in (b). Traveling waves are stable in the region enclosed by the solid, dotted, and dashed lines. Dash-dotted line: neutral curve; dashed line: Eckhaus stability limit. The solid line in (a) and (b), the dotted line in (a), and the dotted line in (b) represent, respectively, three different amplitude instabilities. The open circles in the inset of (b), obtained from the amplitude equations, mark the line of instability due to interaction between oppositely traveling waves. The stability diagram for negative k is a symmetric image with respect to the r-axis. The parameter values are $a = 0.01$, $\sigma = 1$, $\Omega = 0.3$, and $b = 0.1$.

2 Instabilities: Direct Stability Analysis and Phase Equations

To study the stability of the traveling waves, whose traveling direction now can be taken to be in the x-direction, we superimpose infinitesimal disturbance onto them

$$e = e_{tw} + \delta e = e_{tw} + \overline{\delta e}e^{i(kx+\omega t)},$$
$$p = p_{tw} + \delta p = p_{tw} + \overline{\delta p}e^{i(kx+\omega t)}, \tag{9}$$

[1] Standing waves are preferred in the presence of an imperfect O(2) symmetry, see [LMP].

$$n = n_{tw} + \delta n.$$

We have included the factor $e^{i(kx+\omega t)}$ in δe and δp, because in this way we can eliminate this factor from the Maxwell-Bloch equations and obtain a set of linear partial differential equations with constant coefficients after linearization with respect to $\overline{\delta e}$, $\overline{\delta p}$, and δn. This linear stability problem is solved by taking the following *Ansatz* for the disturbances:

$$
\begin{aligned}
\overline{\delta e} &= v_1 e^{i\mathbf{S}\cdot\mathbf{X}} + v_2 e^{-i\mathbf{S}\cdot\mathbf{X}}, \\
\overline{\delta p} &= v_3 e^{i\mathbf{S}\cdot\mathbf{X}} + v_4 e^{-i\mathbf{S}\cdot\mathbf{X}}, \\
\delta n &= v_5 e^{i\mathbf{S}\cdot\mathbf{X}} + v_5^* e^{-i\mathbf{S}\cdot\mathbf{X}},
\end{aligned}
\tag{10}
$$

with $\mathbf{s} = (s_{\mathbf{x}}, s_{\mathbf{y}})$. One is left with a set of five independent ordinary differential equations

$$
\begin{pmatrix}
c_1 + \partial_t & 0 & -\sigma & 0 & -i\delta_1\bar{e} \\
0 & c_2 + \partial_t & 0 & -\sigma & i\delta_1\bar{e} \\
c_{31} & -i\delta_2\bar{e}\bar{p} & c_+ + \partial_t & 0 & \bar{e} \\
i\delta_2\bar{e}\bar{p}^* & c_{42} & 0 & c_- + \partial_t & \bar{e} \\
-\bar{p}^*/2 & -\bar{p}/2 & -\bar{e}/2 & -\bar{e}/2 & b + \partial_t
\end{pmatrix}
\begin{pmatrix}
v_1 \\ v_2^* \\ v_3 \\ v_4^* \\ v_5
\end{pmatrix}
= 0,
\tag{11}
$$

where
$c_1 = i(\omega + ak^2 + as^2 + 2aks_x - \delta_1\bar{n}) + \sigma,$
$c_2 = i(-\omega - ak^2 - as^2 + 2aks_x + \delta_1\bar{n}) + \sigma,$
$c_\pm = 1 \pm i(\Omega + \omega) \mp i\delta_2\bar{e}^2$, $c_{31} = \bar{n} - r - i\delta_2\bar{e}\bar{p}$, and $c_{42} = \bar{n} - r + i\delta_2\bar{e}\bar{p}^*$.
The time dependence of $(v_1, v_2^*, v_3, v_4^*, v_5)^T$ is chosen proportional to $e^{\lambda t}$. Then, we obtain a linear eigen problem with the eigenvalue λ. This system is solved numerically. If for a given k and r, a modulation wavevector $\mathbf{s}$ exists, such that $Re[\lambda(k, r; \mathbf{s})] > 0$, the traveling wave represented by (k, r) is unstable; otherwise, it will be regarded as linearly stable.

The translational symmetry of the problem ensures the existence of a neutral mode at $\mathbf{s} = \mathbf{0}$, and we identify long wavelength unstable growth bands emanating from this neutral stable point as phase instabilities. The most common of these is the Eckhaus instability [Eck], which occurs for $(s_x \to 0, s_y = 0)$, and the zig-zag instability [NPL, Bus], which occurs for $(s_x = 0, s_y \to 0)$. Amplitude instabilities occur at shorter wavelengths, i.e., $\mathbf{s}$ is finite, and may or may not connect directly to the neutral mode at $\mathbf{s} = \mathbf{0}$.

A long wavelength instability can manifest itself as a change in the local wavevector. A description of a local wave pattern and the stability properties are conveniently captured by the Cross-Newell phase equation [CrN]. One of the advantages of this equation is that the rotational degeneracy in the x-y plane (i.e., the wavevector can point in any direction) is built into the formalism from the very beginning, so that the theory is capable of describing large-scale changes in the orientation and wavelength of waves involved in complicated patterns both near and far above threshold [CrN]. Here, it will be used only to derive analytical expressions for various phase-instability boundaries. Let $(e, p, n) = (\bar{e}e^{i\theta}, \bar{p}e^{i\theta}, \bar{n})$. We investigate the stability of the transverse traveling wave, whose wavevector

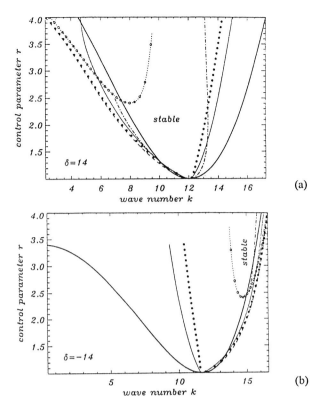

Fig. 2. Stability diagrams for Raman lasers. Thick solid line: neutral curve; dashed line: existence curve of the traveling-wave solution; thin solid line: Eckhaus instability from the phase equation; solid circles: Eckhaus instability from the direct stability analysis; dotted line: zig-zag instability from the phase equation; open circles: zig-zag instability from the direct stability analysis; dash-dotted line: amplitude instability. The fixed parameters are $a = 0.05$, $\sigma = 3$, $\Delta_s = -7$, $b = 0.83$ $\delta_1' = -90$, $\delta_2' = 15$, and $\delta_3' = 1/60$.

$\mathbf{k} = \nabla\theta$ and frequency $\omega = \partial_t\theta$ are fixed constants, by allowing the gradient of the phase θ (local wavevector) to become a slowly varying function of space and time. The equation governing the phase evolution is [Leg]

$$\partial_t\theta = \omega + D_\parallel \partial_x^2\theta + D_\perp \partial_y^2\theta, \tag{12}$$

where

$$D_\parallel = -\frac{1}{\tau(k)}\left[x_1\frac{2ak^2}{\bar{e}}\frac{d\bar{e}^2}{dk^2} + a\bar{e}x_1 + 4s(k)k^2\frac{d\omega}{dk^2}\right], \quad D_\perp = -\frac{a\bar{e}}{\tau(k)}x_1 \tag{13}$$

with

$$x_1 = \frac{2\bar{e}}{\sigma}\left[\alpha + (b\delta_2 - \frac{\delta_1}{\sigma})\alpha^2 - (b\delta_2 + \frac{\delta_1}{\sigma}) - \delta_2\bar{e}^2(1 - 2\delta_1\frac{\alpha}{\sigma})\right],$$

$$\tau(k) = -\frac{2\bar{e}^2}{\sigma}\left[1 + \sigma + 2\alpha(b\delta_2 - \delta_1)\right],$$

$$s(k) = \frac{d\bar{e}^2}{dk^2}\left[\frac{\alpha}{\sigma} + (b\delta_2 - \frac{\delta_1}{\sigma})\frac{\alpha^2}{\sigma} - \frac{1}{\sigma}(b\delta_2 + \frac{\delta_1}{\sigma})\right.$$

$$-\frac{\delta_2}{\sigma}(1 - 2\delta_1\frac{\alpha}{\sigma})\bar{e}^2 + \frac{2}{b}\delta_2\bar{e}^2(1 - 2\delta_1\frac{\alpha}{\sigma})$$

$$-(b\delta_2 + \frac{\delta_1}{\sigma}) - \frac{3\bar{e}^2}{\sigma + 1}(\delta_2 - \frac{\delta_1}{\sigma})(1 + \alpha(b\delta_2 - \frac{\delta_1}{\sigma})) +$$

$$\left.\frac{\Omega - ak^2}{1 + \sigma}(1 + \alpha(b\delta_2 - \frac{\delta_1}{\sigma}))\right] - \frac{2a\bar{e}}{\sigma + 1}\left[1 + \alpha(b\delta_2 - \frac{\delta_1}{\sigma})\right].$$

The Eckhaus and zig-zag instability boundaries are given by
$D_\parallel = 0$ and $D_\perp = 0$ ($D_{\parallel(\perp)} < 0 \rightarrow$ Eckhaus and (zig-zag) unstable, respectively).

In the following, we give the results of the stability analysis. For the two-level laser, we shall concentrate on the 1-D situation (no y-dependence), which can be achieved in principle by using an appropriate annular geometry. We state straightaway that the only two-dimensional instability we found is the zig-zag instability, which destabilizes all traveling waves whose wave number k is less than k_c. Figure 1a is the stability diagram in the (k, r)-plane for a two-level laser. The near-threshold regime and its local enlargement is shown in Fig. 1b. The stable region of traveling-wave solutions is bounded by the Eckhaus instability (dashed line) and instabilities (solid and dotted lines), which we call *amplitude instabilities*, because they involve the excitation of new modes with distinctly different wave numbers, i.e., s ($\mathbf{s} = (\mathbf{s}, \mathbf{0})$) is finite. The Eckhaus curves obtained from the direct stability analysis and from the phase equation coincide exactly. The amplitude instabilities can only be calculated from the direct stability analysis. The amplitude instabilities are denoted by the solid line in Fig. 1a, b, the dotted line in Fig. 1a, and the dotted line in Fig. 1b. Each sets in at values of the perturbation wave number s different from zero and with finite values of $Im(\lambda)$, which are of the same order of magnitude as the frequencies of the traveling waves. At the dotted line in Fig. 1a, where an amplitude instability sets in, the most unstable perturbation wave number s_m, corresponding to maximum growth rate $max_s Re(\lambda)$, and is typically of order k_c, but varies along the stability border. From the point of view of experimentalists, this instability may be of limited interest, because it occurs for values of r too far above threshold. The dotted line in Fig. 1b marks another amplitude instability. As (k, r) move along this instability line toward the Eckhaus border, s_m approaches zero and the growth-rate curve merges into the Eckhaus growth-rate curve.

The amplitude instability that sets in at the solid lines of Fig. 1a, b is traced back to the interaction between oppositely traveling waves. Here the most excited mode is s_m near $2k_c$. Near threshold it lies in the Eckhaus unstable regime, and one observes that the perturbation wave amplitudes v_2 and v_4 dominate v_1, v_3, and v_5, and that $s_m \approx 2k_c$, $Im(\lambda) \approx 0$. So the instability gives rise to a left-traveling wave with wave number near $2k_c - k$ (see (9, 10)). (The instability is analogous to the cross-roll instability [Bus] experienced by stationary convection rolls at a large Prandtl number in which, as the wave number is increased, the

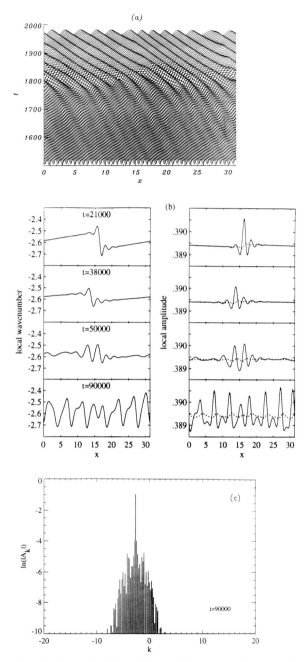

Fig. 3. Numerical simulation with an unstable traveling wave of $k = 9, r = 2.53$ as the initial state (the growth-rate curve is given by open circles in Fig. 2. (a) Transition from right-traveling wave to left-traveling wave (real part of the electric field *versus* (x, t)). (b) Local wave number at different times (left half) and amplitude of the electric field (solid lines) at corresponding times (right half). The dotted lines are the local amplitude evaluated from the local wave number. (c) Spatio spectrum at $t = 90,000$.

pattern develops an instability to an orthogonal set of rolls with a wave number close to k_c.) We have confirmed this picture by looking at the interaction between oppositely traveling waves near threshold for which the amplitude equations are [Jk1, NwM]

$$\tau_0(\partial_t + v_g\partial_x)A_1 = \epsilon A_1 + \xi_0^2(1 + ic_1)\partial_x^2 A_1 - g(1 + ic_2)(|A_1|^2 + \beta|A_2|^2)A_1,$$
$$\tau_0(\partial_t - v_g\partial_x)A_2 = \epsilon A_2 + \xi_0^2(1 + ic_1)\partial_x^2 A_2 - g(1 + ic_2)(|A_2|^2 + \beta|A_1|^2)A_2,$$

where $\tau_0 = (\sigma + 1)/\sigma$, $\epsilon = r - 1$, $v_g = 2ak_c/(\sigma + 1) = \xi_0$, $c_1 = (1 + \sigma)^2/(4\sigma a k_c^2)$, $g = 1/b$, and $\beta = 2$. A_1 and A_2 are the amplitudes of right- and left-traveling waves, respectively

$$(e, p, n) = \boldsymbol{U}\left[A_1 e^{i(k_c x - \omega_c t)} + A_2 e^{-i(k_c x + \omega_c t)}\right].$$

($\boldsymbol{U} = (1, 1 - i\Omega/(\sigma + 1), 0)$ is the linear eigenvector.)

From the amplitude equations, one can solve for the right-traveling-wave solution $A_1 = \rho e^{i(qx - \nu t)}$ with $\rho = \sqrt{(\epsilon - \xi_0^2 q^2)/g}$, (note that the wave number is $k = k_c + q$), then perturb it with a small left-traveling wave $A_2 \propto e^{\gamma t + ipx}$ (wave number $= -k_c + p$). Stability requires $\max_p Re(\gamma(p)) < 0$, which is

$$(k - k_c)^2 = q^2 < (1 - \frac{1}{\beta})\frac{\epsilon}{\xi_0^2} = \frac{1}{2}\left(\frac{\sigma + 1}{2ak_c}\right)^2 (r - 1). \qquad (14)$$

The stability boundary is shown by the open circles in the inset of Fig. 1b, and it agrees well with the numerical results for r close to $r_c = 1$. From (14) one sees that this instability happens only for $\beta > 1$, the regime in which traveling waves are preferred to standing waves. This means that the interaction between right- and left-traveling waves must be strong enough for the instability to occur. In our case, $\beta = 2$ and is parameter independent. The extension of the instability curve (14) to higher values of r is the solid line in Fig. 1.

Figure 2 gives stability diagrams for Raman lasers. Again, we find the Eckhaus instability, zig-zag instability, and amplitude instabilities, which set in for $s_y = 0$. Another important type of phase instability, Benjamin-Feir instability, can also occur for a different choice of the parameter set [Leg]. Compared to two-level lasers, the Raman laser exhibits several interesting features. First, traveling-wave solutions can exist in a region below the neutral curve (solid line) and above the existence curve (dashed lines), therefore, we have subcritical bifurcation. Second, the stable region for $\delta = -14$ is shifted to a higher r and k region. Notably, there are no stable traveling-wave solutions for $k = k_c$. Third, while the agreement between the zig-zag instability boundaries from the direct stability analysis (open circles) and from the phase equation (dotted lines) is very good, there is a large discrepancy on the Eckhaus instability between the two methods. The correct Eckhaus instability boundary is the one obtained from the direct stability analysis. The discrepancy, possibly due to the interaction between the Eckhaus and amplitude instabilities, and the interesting phase instability (zig-zag and Eckhaus) behavior have yet to be explored.

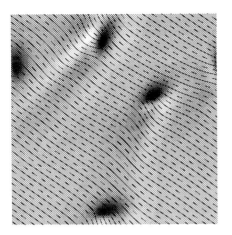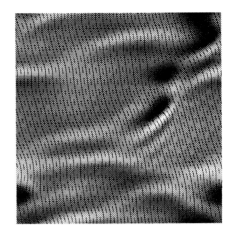

Fig. 4. Snapshots of 2-D wave patterns of the Raman laser for $\delta = 14$ (left) $\delta = -14$ (right), starting from random initial conditions (other parameters are the same as in Fig. 2). The amplitude of the electric field is reflected by the gray scale and the black lines (continuous and dashed lines) are the iso-phase lines.

3 Pattern Transition and Selection

Pattern transition and selection are related topics. We now use a 1-D, two-level laser to illustrate the transition process from a traveling wave to a chaotically modulated traveling wave when an amplitude instability sets in, and we use a 2-D Raman laser to show how patterns may develop across the laser cross section when the stress parameter r is chosen, so that Eckhaus, zig-zag, amplitude unstable, and stable traveling-wave solutions coexist.

The presence of the amplitude instability due to the interaction between oppositely traveling waves severely restricts the stability domain. To see what state the system will evolve to when this instability sets in, we solve (1) numerically for a 1-D, two-level laser with an unstable traveling wave of $(k = 9, r = 2.53)$ as the initial condition, with a very small perturbation added to it. The boundary condition in the transverse x-direction is chosen to be periodic. The initial traveling wave is unstable with respect to a band of perturbations: $Re(\lambda) > 0$ for $8.5 < s < 13.1$ and $s_m = 11.2$. We follow the evolution in Fig. 3. Fig. 3a shows a dramatic transition from a right-traveling wave to a left-traveling wave. One can see that the wavelength of the left-going wave is not uniform in space (see also Fig. 3b), because there are large spatial intervals where it changes slowly. This slowly varying feature can be captured by the *local wave number*, which is defined as a gradient of the phase of the complex electric field envelope

$$e(x, t) \equiv A(x, t) = |A(x, t)|e^{i\theta(x, t)},$$

$$k_{loc} \equiv \partial_x \theta = [Re(A)\partial_x Im(A) - Im(A)\partial_x Re(A)]/|A|^2.$$

Figure 3b illustrates the evolution of the local wave number (left half) of the resulting left-traveling wave. On the right half, the solid lines give the local amplitude $|A|$ at corresponding times, and the dotted lines are obtained from $\sqrt{b[r - r_0(k_{loc})]}$, the formula for the amplitude of pure traveling waves (see (3)). At time $t = 21,000$, there is a narrow interval in which the local wave number and amplitude change rapidly, embedded in a large interval, where k_{loc} changes slowly, in fact, linearly giving rise to an overall chirp in the phase, and where the local amplitude evaluated from k_{loc} agrees remarkably well with the exact amplitude. This indicates that the phase is the only active variable and that the amplitude is slaved to the phase. The narrow, fast-varying region also travels to the left. As time goes on, it expands into the surrounding slowly varying domain, and the slope of k_{loc} in the slowly varying domain decreases ($t = 38,000$). As k_{loc} in the slowly varying domain becomes more and more uniform, one can expect that the Eckhaus instability will occur, because the central wave number is in the Eckhaus unstable region (see Fig. 1). The oscillation in the local wave number and amplitude at positions away from the fast varying domain for $t = 50,000$ appears to be generated by the Eckhaus instability. Eventually, the system evolves into a weakly turbulent state ($t = 90,000$) in the sense that amplitude modulation is spatio-temporally chaotic, but very small. In this state, the amplitude no longer follows the phase everywhere, as indicated by the discrepancy between solid and dotted lines for $t = 90,000$, and is itself an active variable. We show in Fig. 3c the wave-number spectrum for this weakly turbulent state. It is dominated by a wave with wave number $k_f = -2.6$, which is close to the most unstable wave number $k - s_m = 9 - 11.2 = -2.2$ of the initial wave. (In another run with the size of x-domain doubled, we obtained $k_f = -2.15$.)

We have also carried out a direct numerical simulation of an Eckhaus unstable wave with wave number k_f and observed the direct transition to turbulent states very similar to the one described above.

Figure 4 shows typical wave patterns in the x-y cross sections of Raman lasers. Here the pump is chosen so that the Eckhaus, zig-zag, amplitude unstable, and stable Busse balloon coexist. The field is initiated from noise. The pictures show finite amplitude wave patterns, consisting of a sea of optical vortices (zeros of the complex field), and bright ridges (maxima of the field), which appear to arise from a combination of Eckhaus, zig-zag, and amplitude instabilities. The bright ridges are aligned roughly at right angles to the local direction of the traveling waves. This complicated spatio-temporal evolution appears to persist indefinitely with no sign of any regular recurrence.

4 Conclusion

In this survey, we have shown that both two-level and Raman wide-aperture lasers are capable of displaying a rich variety of pattern-forming instabilities. The relative simplicity of the exact nontrivial lasing solution for these models affords a unique opportunity to evaluate the validity of the universal phase equation and amplitude equation descriptions in a physically important context. The analysis can be extended to include the effects of inhomogeneous broadening and multi-longitudinal mode oscillations, and it is anticipated that much of the spatio-temporal behavior of two-level and Raman lasers will carry over to techno-logically important wide-aperture semiconductor laser systems, which require a much more complicated material description, involving many-body interactions between carriers and holes at the microscopic level. We expect that the novel amplitude instability of laser systems is accessible in many other contexts, such as convection in binary mixture fluids and liquid crystals, which show many of the symmetries and characteristics of the laser optics situation.

Acknowledgments The authors wish to thank the Arizona Center for Mathematical Sciences (ACMS) for support. ACMS is sponsored by AFOSR contract FQ8671-9000589 (AFOSR-90-0021). Two of the authors (J.V.M. and J.L.) acknowledge partial support from a European Community Twinning grant SCI*0325-C(SMA).

References

[NPL]   Newell, A.C., Passot, T., Lega, J.: Order Parameter Equations for Patterns. Ann. Rev. Fluid Mech. **25** 1993, 399.

[AbF]   Abraham, N.B., Firth, W.J.: J. Opt. Soc. Am. B **7**, 1990, 951, and the references therein. For more recent references, see, e.g., Arecchi, F.T., Boccaletti, S., Mindlin, G.B., Pérez García, C.: Phys. Rev. Lett. **69** 1992, 3723; D'Angelo, E.J., Green, C., Tredicce, J.R., Abraham, N.B., Balle, S., Chen, Z., Oppo, G.L.: Physica D **61** 1992, 6; Arecchi, F.T., Boccaletti, S., Giacomelli, G., Puccioni, G.P., Ramazza, P.L., Residori, S.: *ibid* **61** 1992, 25.

[CGR]   Coullet, P., Gil, L., Rocca, F.: Opt. Comm. **73** 1989, 403.

[Jk1]   Jakobsen, P.K., Moloney, J.V., Newell, A.C., Indik, R.: Phys. Rev. A **45** 1992, 8129.

[NwM]   Newell, A.C., Moloney, J.V.: Nonlinear Optics. Addison-Wesley, Redwood City, CA, 1992.

[Bus]   Busse, F.H.: Rep. Prog. Phys. **41** 1978, 28.

[CrN]   Cross, M.C., Newell, A.C.: Physica D **10**, 1984, 299.

[NPS]   Newell, A.C., Passot, T., Souli, M. J.: Fluid Mech., **220** 1990, 187.

[Leg]   Lega, J., Jakobsen, P.K., Moloney, J.V., Newell, A.C.: Nonlinear Transverse Modes of Large-Aspect-Ratio Homogeneously Broadened Lasers: II. Pattern Analysis Near and Beyond Threshold. Phys. Rev. A. **49** (5) 1994, 4201.

[PpR]   Paap, H.-G., Riecke, H.: Phys. Rev. A **41** 1990, 1943.

[FMN]   Feng, Q., Moloney, J.V., Newell, A.C.: Amplitude Instabilities of Transverse Traveling Waves in Lasers. Phys. Rev. Lett. 71 **11** 1993, 1705 - 1708.

[Jk2]   Jakobsen, P.K., Lega, J., Feng, Q., Staley, M., Moloney, J.V., Newell, A.C.: Nonlinear Transverse Modes of Large-Aspect-Ratio Homogeneously Broadened Lasers: I. Analysis and Numerical Simulation. Phys. Rev. A **5** (49) 1994, 4189.

[LMP]   López Ruiz, R., Mindlin, G.B., Pérez García, C.: Phys. Rev. A **47** 1993, 500.

[Eck]   Eckhaus, W.: Studies in Nonlinear Stability Theory. Springer, New York, 1965; Kramer, L., and Zimmermann, W.: Physica D **16** 1985, 221; Janiaud, B., Pumir, A., Bensimon, D., Croquette, V., Richter, H., Kramer, L.: Physica D **55** 1992, 269.

6 Laser-Based Optical Associative Memories

M. Brambilla,[1,2] L. A. Lugiato,[1] M. V. Pinna,[1] F. Pratti,[1,2] P. Pagani,[3] and P. Vanotti[3]

[1] Dipartimento di Fisica dell' Università, Via Celoria 16, 20133 Milano, Italy
[2] Physik-Institut der Universität Zürich, Switzerland
[3] Alenia S.p.A., viale Europa 1, 20014 Nerviano, Italy

Introduction

A new approach to architecture based on the use of a laser as a nonlinear discriminator for all-optical, auto- and hetero-associative memories is presented. The laser operates in a regime of spatial multistability, i.e., of the coexistence of different stable stationary states. Numerical calculations in the simplest situations indicate that the laser is able to decide which of its stationary states is most similar to a field pattern that is injected into the laser itself. The laser is combined with a linear system composed of lenses, holograms, and a pinhole mask. The memory is constituted of a certain number of images that are stored in one of the holograms and which are in one-to-one correspondence with the stationary states of the laser. The task of the linear part is (a) to convert an arbitrary image offered to the system into an appropriate field pattern, which is injected into the laser; and (b) to convert the stationary beam, which emerges from the laser, into the corresponding image in memory.

During recent years, a general interest has grown in several fields of the applied sciences in an unconventional calculus based on analogies with neural activities. An important feature of this approach is a memory element whose address is made by association of contents rather than by position. This kind of memory is well suited for an optical implementation due to its high density of interconnections [Gdm].

The first application of optical properties to associative memories by Gabor dates back to 1969 [Gbr]; also, the use of devices that realize the correlation between images on corresponding Van der Lugt matched filters [VdL] represented an important step in this direction. However, only in recent years have there been demonstrations, thanks to the introduction of nonlinear devices, which accomplish a threshold function and permit a drastic reduction of the signal-to-noise ratio in the output.

Numerical algorithms for creating and operating an associative memory were formulated by Hopfield [Hph], Kohonen [Khn], and Grossberg [CrG]. Cohen· [Chn] proposed a holographic model for the realization of neural networks. The first example of an optical associative memory based on the Hopfield neural network model was reported by Psaltis and Farhat in 1985 [Ps1]. Major experimental works are due to Psaltis and collaborators [Ps2], who for the first time used a nonlinear element in an optical loop; to Soffer and colleagues [Sfr]; to Yariv and colleagues [YrK]; to Anderson and colleagues [ALF], who demonstrated recall and processing of images using photorefractive crystals; and to Lalanne and colleagues [LCT], who realized an optical inner-product implementation of the Hopfield model.

We started an investigation intended to explore the possibility of using the laser as the nonlinear element in an associative memory architecture. This possibility was first suggested by Haken [Hkn, FuH], and our research has been inspired by his formalization of synergetic computers for pattern recognition and associative memories. Our approach is also in the line indicated by Anderson [And], who suggested the opportunity of utilizing continuous field distributions rather than discrete arrays of elements and emphasized the importance of "mode" competition for the associative memory operation. The investigations of Vorontsov and collaborators [Vrn], utilizing liquid crystals in an optical architecture with feedback, are also in this line.

The starting point of our approach is the effect of "spatial multistability," which was predicted and experimentally verified in our previous research on transverse laser patterns [TmW, Br1, Br2]. This phenomenon can be described in the simplest way by saying that under appropriate conditions for the parameters of the system, when the laser is switched above threshold, it can approach different steady states, which differ for the transverse structure of the emitted beam. When the system is isolated, the choice of the final stationary state is completely random, because an initial fluctuation determines the evolution of the system.

Very different is the behavior when the laser is driven by an external coherent beam with the same frequency, so that the system has the configuration of a laser with injected signal. The system is operated in precisely the following way. In the initial stage, the laser is below threshold and a weak signal is injected into the laser cavity. In the second stage, the laser is switched above threshold, while the signal is switched off (the second operation is not strictly necessary, in the sense that it does not influence the following behavior of the system). In this case, the evolution of the laser starts from the seed left by the signal, so that the choice of the stationary state is no longer random.

We analyzed the validity of a simple criterion to predict the selection of the steady state as a function of the spatial configuration of the signal. This criterion is based on the introduction of an appropriate "similarity parameter." Precisely, two transverse field configurations are compared on the basis of their inner product, which coincides with the correlation function of the two patterns evaluated on the laser axis. The criterion states that the laser selects the stationary state that is most similar to the field pattern of the injected signal.

Our calculations show that in the largest majority of cases the selection of the final stationary state agrees with the similarity criterion. The exceptions correspond to an ambiguous situation in which two different stationary states display similarity parameters that differ by a few percent.

These results prove that the laser works as a *decision element*, in the sense that it is able to decide which of the stationary states is most similar to the input signal. In turn, this capability suggests the feasibility of this system for associative memory/pattern recognition operations; in this case, the memory of the system is constituted by the stationary states.

If one wants a standard associative memory in which the "objects" in memory are ordinary and arbitrary images, it is necessary to complement the laser with another system that establishes a one-to-one correspondence between ordinary images in memory I_n and stationary laser patterns B_n. Different from the laser, this complementary system is linear and consists of a suitable arrangement of lenses, two holograms (one of which is computer generated [BrL, Lee]), and pinholes. The architecture of the linear system is identical to that considered in [LCT] (see also [Ps1, Ps2]), but its characteristic properties are specific and tailored for its use in connection with the laser as the discriminator element. In our scheme, the images in memory are recorded in one of the holograms and the system is reconfigurable in the sense that the change of the set of images in memory does not require any change in the other elements of the system.

An arbitrary image I, which is offered to the system (it is, for example, a partial or a corrupted version of one of the images in memory, say $I_{\overline{n}}$)), is converted by the linear system into a corresponding field configuration A, which is injected into the laser (Fig. 1). The system is devised in such a way that the values of the similarity parameters for I, when they are compared with the images in memory I_n, are proportional, respectively, to the values of the similarity parameter for A compared to the corresponding stationary state of the laser B_n. As a consequence, the laser approaches the stationary state $B_{\overline{n}}$, which corresponds to $I_{\overline{n}}$. The output of the laser, i.e., the stationary field configuration $B_{\overline{n}}$, propagates then in the backward direction through the linear system and produces the output of the complete system (Fig. 1). The second relevant characteristic of the linear system is that, in the backward direction, it converts each stationary state B_n into the corresponding image I_n. In conclusion, the system as a whole (linear + nonlinear) operates in such a way that the input image I is transformed into the image in memory $I_{\overline{n}}$, which is most similar to I.

1 Nonlinear Dynamic Equations and Steady-State Equations

We consider a ring laser with spherical mirrors and assume that the length of the active region is much smaller than the Rayleigh length of the cavity. This circumstance allows us to neglect the longitudinal variations of the beam width and field phase along the active sample. Actually, this restriction is not essential for the nature of the results that we will derive and can be easily dropped; however, we prefer to use it to keep our calculations as simple as possible. Hence, for a cylindrically symmetrical cavity, the transverse profile of the cavity modes

Optical Associative Memory: basic system

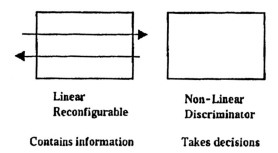

Linear
Reconfigurable

Non-Linear
Discriminator

Contains information

Takes decisions

Fig. 1. Scheme for a laser-based optical associative memory.

is described by the functions [Kgl]

$$A_{pl}(\rho, \varphi) = \sqrt{\frac{2}{\pi}} (2\rho^2)^{|l|/2} L_p^{|l|}(2\rho^2) e^{-\rho^2} e^{il\varphi}, \qquad (1)$$

where $p = 0, 1, ...$ is the radial index and $l = 0, \pm 1, \pm 2, ...$ is the angular index, ρ denotes the radial coordinate $r = (x^2 + y^2)^{1/2}$ normalized to the beam waist w (x and y are the Cartesian coordinates in the transverse plane), φ is the angular coordinate, and L_p^l are Laguerre polynomials of the indicated argument. The functions A_{pl} obey the orthonormality relation

$$\int_0^{2\pi} d\varphi \int_0^\infty d\rho\, \rho\, A_{pl}(\rho, \varphi)\, A_{p'l'}^*(\rho, \varphi) = \delta_{pp'}\, \delta_{ll'}. \qquad (2)$$

We assume that the active medium is a homogeneously broadened system of two-level atoms with transition frequency ω_a and line width $\gamma_\perp$, and that the excited region has a Gaussian transverse shape of radius r_p, i.e., the transverse configuration of the equilibrium population inversion is described by the function

$$\chi(\rho) = \exp(-2\rho^2/\psi^2), \qquad \psi = 2r_p/w. \qquad (3)$$

An important property of the Gauss-Laguerre modes is that their frequency depends on the transversal mode indices p and l via the combination $2p + |l|$, a situation that produces mode degeneracy. We assume now that the atomic line is in resonance with a frequency-degenerate family of transverse modes corresponding to a given value of the longitudinal modal index, such that $2p+|l| = q$, with q fixed. We call ω_q the common frequency of the modes of the family. We suppose, in addition, that all the other cavity modes either suffer from large losses or that their frequency separation from the atomic line is much larger than the atomic line width; therefore, only the modes belonging to the frequency-degenerate family take part in the laser emission. Hence, the laser field has the form

$$E(\rho, \varphi, z, t) \propto F(\rho, \varphi, z, t) \exp(-\imath\omega_q t + \imath k_q z) + c.c., \qquad (4)$$

where F is the normalized slowly varying envelope and $k_q = \omega_q/c$ is the common wave number of the modes of the family. In turn, the envelope F can be expanded as follows:

$$F(\rho, \varphi, t) = {\sum_{pl}}' f_{pl}(t) A_{pl}(\rho, \varphi), \qquad (5)$$

where the prime indicates that the sum is restricted to the modes of the family. The modal amplitudes f_{pl} obey the time evolution equations [Lug]

$$\frac{df_{pl}}{dt} = -k\left[f_{pl} - 2C \int_0^{2\pi} d\varphi \int_0^\infty d\rho \, \rho A_{pl}^*(\rho, \varphi) P(\rho, \varphi, t) \right], \qquad (6)$$

where k is the cavity damping constant

$$k = \frac{cT}{\mathcal{L}}, \qquad (7)$$

T is the transmissivity coefficient of the mirrors, $\mathcal{L}$ is the total length of the ring cavity, and C is the pump parameter

$$C = \frac{\alpha L}{2T}, \qquad (8)$$

α is the gain parameter of the field per unit length and L is the length of the active region; P is the normalized slowly varying envelope of the atomic polarization. Equation (6) must be coupled with the atomic Bloch equations, which read

$$\frac{\partial P}{\partial t} = \gamma_\perp \left[F(\rho, \varphi, t) D(\rho, \varphi, t) - P(\rho, \varphi, t) \right], \qquad (9)$$

$$\frac{\partial D}{\partial t} = -\gamma_\| \left[\mathcal{R}e \left(F^*(\rho, \varphi, t) P(\rho, \varphi, t) \right) + D(\rho, \varphi, t) - \chi(\rho) \right], \qquad (10)$$

where D is the normalized population inversion and $\gamma_\|$ is its relaxation rate. In (9) and (10), it is understood that F is given by the expansion (5); thus, (6), (9), and (10) form a self-contained set of equations for the variables f_{pl}, P, and D.

In the steady state, one obtains from (9) and (10)

$$D = \frac{1}{1 + |F(\rho, \varphi)|^2} \chi(\rho), \qquad P = \frac{F(\rho, \varphi)}{1 + |F(\rho, \varphi)|^2} \chi(\rho), \qquad (11)$$

so that (6) becomes

$$f_{pl} = 2C {\sum_{p'l'}}' \int_0^{2\pi} d\varphi \int_0^\infty d\rho \, \rho \frac{A_{pl}^*(\rho, \varphi) A_{p'l'}(\rho, \varphi)}{1 + |F(\rho, \varphi)|^2} \chi(\rho) f_{p'l'}. \qquad (12)$$

Of course, (12) admits the trivial stationary solution $f_{pl} = 0$ for all the choices of the indices p, l in the family.

2 Single- and Multimode Stationary Solutions. Spatial Multistability

The set of stationary equations (12) also admits exact single-mode solutions, in each of which only one transverse mode is excited, while the amplitudes of all the other modes are exactly equal to zero.

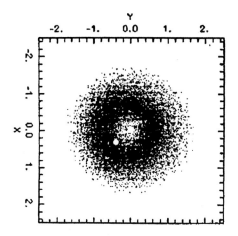

Fig. 2. Intensity distribution for the doughnut modes $p = 0$, $l = \pm 1$.

For example, in the case of $2p + |l| = 1$, the modes of the empty cavity are the two doughnut modes $A_{0,\pm 1}$, whose form is

$$A_{0,\pm 1} = \sqrt{\frac{2}{\pi}} \sqrt{2\rho^2} e^{-\rho^2} e^{\pm i\varphi} . \tag{13}$$

The intensity distribution of these modes is the same for both modes and is shown in Fig. 2. For this family, all the steady states have the form (13), up to a proportional constant. The stability analysis shows that these single-mode steady states are stable for all values of the pump parameter. This implies that these two steady states coexist and that there is spatial bistability between them; this phenomenon was observed experimentally in [TmW].

On the other hand, in general, the steady-state equations (12) admit also multimode solutions, i.e., states to which several modes of the family contribute by locking their phases in appropriate ways. This is already evident in the case of the frequency-degenerate family. Figure 3 shows all the stable stationary patterns that one finds by varying the control parameters C and Ψ, in the case of the degenerate family $2p + |l| = 2$.

Figures 3a and 3d correspond to the single-mode states $p = 1, l = 0$ and $p = 1, l = \pm 2$, respectively (the two states $p = 1, l = +2$, and $p = 1, l = -2$ have the same transverse intensity distribution). Patterns shown in Figs. 3b and 3c

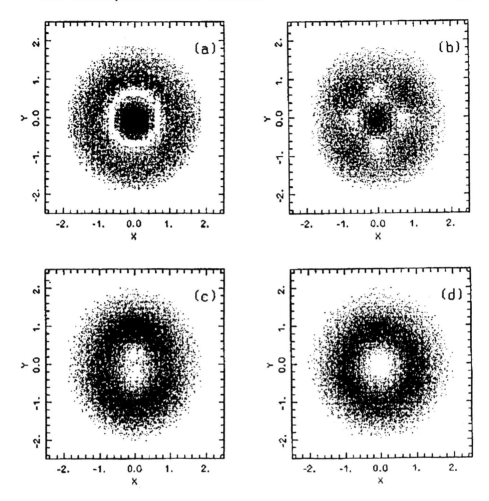

Fig. 3. Intensity distribution of the four stationary patterns for the case $2p + |l| = 2$. (a) Single-mode pattern $p = 0$, $l = 0$, (b) "Four-holes," (c) "Oval," (d) Doughnut $p = 0$, $l = \pm 2$.

correspond to multimode stationary solutions in which all of the three modes $p = 1, l = 0$, and $p = 1, l = \pm 2$ are present with different weights. Due to its shape, we call "oval" the state shown in Fig. 3c, while we designate by "four holes" the pattern in Fig. 3b, because it displays four points in which the intensity is exactly zero. Actually, these points are centers of optical vortices [Br1]; the configurations of Fig. 3c and 3d show, instead, two and one vortex, respectively.

As it is shown in [Br1], in the plane (C, Ψ) of the control parameters there is a sizeable domain in which the four-hole pattern coexists with the two single-doughnut-mode solutions; similarly, there is another domain in which the four-hole pattern coexists with the oval solutions, which are obtained from each other by complex conjugation. Both cases are examples of spatial multistability.

3 Operation with Injected Signal

We wish to explore the possibility of employing the laser as a device in which one can store information and which is able to recognize an external "object" on the basis of its memory. Such a device should be able to perform an associative memory function; strictly, this means that when the system is given a partial or ambiguous representation of a memory-stored pattern, it is capable of "remembering" (reconstructing) the whole object. This process of reconstruction implies that the system is capable of "deciding" which among the objects stored in its memory has the maximum similarity to the proposed pattern, hence the system must enhance the dominant features of the ambiguous input pattern and delete all other elements. We do not intend to introduce a general concept of similarity, including, for example, scale invariance, but we simply state that two patterns are most similar when they are superimposable.

A natural definition of the similarity parameter is precisely the following. If $A(\rho, \varphi)$ and $B(\rho, \varphi)$ are the slowly varying envelopes of two transverse configurations of the electric field, we define the similarity parameter by means of the inner product, i.e.,

$$S(A, B) = \frac{\left| \int_0^{2\pi} d\varphi \int_0^\infty d\rho \, \rho \, A(\rho, \varphi) \, B^*(\rho, \varphi) \right|}{\left(\int_0^{2\pi} d\varphi \int_0^\infty d\rho \, \rho \, |A(\rho, \varphi)|^2 \right)^{1/2} \left(\int_0^{2\pi} d\varphi \int_0^\infty d\rho \, \rho \, |B(\rho, \varphi)|^2 \right)^{1/2}}, \quad (14)$$

so that, by the Schwartz inequality, $0 < S \leq 1$ and $S = 1$ when the two patterns A and B coincide (apart from their total intensities). The normalizing factors are introduced to make S independent of the total intensities of the two patterns A and B.

The similarity parameter (14) is immediately related to the correlation function

$$(A \star B)(x', y') \equiv \int_{-\infty}^{+\infty} dx \int_{-\infty}^{+\infty} dy \, A(x, y) \, B^*(x - x', y - y') \quad (15)$$

evaluated at the origin $x' = y' = 0$ (we remind that x and y denote the Cartesian coordinates in the transverse plane); the correlation functions for $x', y' \neq 0$ correspond to the similarity parameter between A and a translated version of pattern B.

In the case that interests us, B_n represents the coexistent stable single- or multimode stationary states of the laser operating in a regime of spatial multistability; these stationary states constitute the memory of our device. On the other hand, A can be thought of as the transverse configuration of an external beam injected into the laser. For example, A might correspond to a stationary state $B_{\bar{n}}$, which has been partially "mutilated" by setting to zero a spatial region of its intensity profile. The associative memory operation requires the laser to recall the whole object by approaching the stationary state $B_{\bar{n}}$.

Our numerical calculations [Br3] show that, in the case of the frequency-degenerate family $2p + |l| = 1$, the laser indeed approaches, and with no errors the doughnut stationary state results in more similarity to the injected field.

In the case of the family $2p + |l| = 2$, we considered several points of the domain in the parameter space, where the "four-hole" stationary pattern coexists with the two doughnut stationary states and several initial conditions.

In almost all cases, the selection of the stationary state agrees with the criterion of similarity with the injected pattern. The few cases of failure correspond to a situation of strong ambiguity (two stationary states with similarity parameters differing by less than five percent).

4 General Description of the System

The results in the preceding section show that the laser itself could be considered as a self-consistent all-optical architecture for association. But, using the same expression as Anderson [An1], this is not an interesting memory, not to anyone whose imagination goes beyond Gauss-Laguerre modes and a simple linear combination thereof. Moreover, once parameters are defined, the memory of this system is fixed and constituted by the stationary laser patterns.

If, on the other hand, one needs an associative memory, operating on generic 2-D images, it is necessary to complete the laser with another architecture that establishes a one-to-one correspondence between ordinary images and stationary states. The whole system, shown in Fig. 1 is, therefore, defined by two distinct parts: the laser, which is nonlinear and makes decisions; and the converter, which is linear and contains information. A generic input image will be converted by means of the linear part into a field shaped on the basis of stationary laser patterns.

The action is performed following a two-step operation. At first, all the features previously stored in a holographic-memory element are extracted, and the degree of presence for each feature is weighted. As a second step, an output seed-field is created as a weighted linear combination of synthetic fields that have to assure a biorthogonality relation with stationary laser patterns.

The weighting operation, as usual [FrP, LCT], is performed by means of a conventional 2-D correlator followed by a pinhole mask, a configuration that allows extraction of the set of inner products between stored patterns and the input image. The correlator can be implemented using a Van der Lugt architecture [VdL] or a Joint Transform Correlator (JTC) based on dynamic holography in photorefractives [RBH]. As shown in Fig. 4, which details the linear part, this is performed by the lenses L_1 and L_2, the hologram H_1, and the pinhole mask.

If we describe with $\{I_n\}$, the set of normalized reference images stored in the hologram H_1 and with $I(x, y)$ the generic input image, the field behind the pinhole mask, is

$$U(x, y) = \sum_{n=1}^{N} a_n\, \delta(x - x_n, y - y_n)\,, \tag{16}$$

L₁   H₁   L₂   P.H.   L₃   H₂

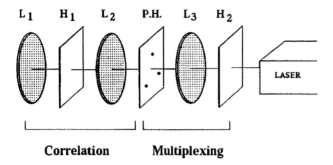

LASER

Correlation   Multiplexing

Fig. 4. Architecture of the linear part of the system, L_1, L_2, L_3 = lenses, H_1, H_2 = holograms, P.H. = pinhole mask.

where

$$a_n = \int\int_{-\infty}^{+\infty} I(x,y)I_n^*(x,y)\, dx\, dy \tag{17}$$

is the inner (scalar) product of $I(x,y)$ with $I_n(x,y)$ (a_n is real, because I and I_n are both real). The number N of stored images coincides with the number of coexisting stationary states of the laser for the chosen operating parameters.

Using the Euclidean representation in a vector space, we can observe that the set $\{i_n\}$ represents a basis (though incomplete) of vectors $|I_n>$. In such a case, the scalar product $< I|I_n> = a_n$ should be viewed as a measure of the degree of presence of the feature I_n in the image I.

The second part of the linear architecture, similar to that of [LCT], acts as a multiplexer of information stored in the hologram H_2 in the sense that each pinhole performs an on-axis recall of the associated field from H_2 (angular multiplexing).

Therefore, the total on-axis field to the right of the hologram is

$$A_0(\xi,\eta) = \sum_{n=1}^{N} c_n W_n(\xi,\eta) = \epsilon \sum_{n=1}^{N} a_n W_n(\xi,\eta), \tag{18}$$

where ϵ represents the total diffraction efficiency (H_1 plus H_2) and $W_n(\xi,\eta)$ is the nth field stored in H_2. The field $A_0(\xi,\eta)$ constitutes the seed that is injected into the laser.

Using the vector-space notation, we can represent the seed field as a vector $|A_0>$ expanded on the basis of the functions $|W_n>$, as indicated in (18). The laser, at first, evaluates all the similarity parameters between $|A_0>$ and its own stationary solutions $|B_i>$ $(i=1,...,N)$ as

$$s_i = \tilde{S}(A_0, B_i) = \frac{< A_0|B_i >}{\|A_0\|\|B_i\|} = \epsilon \sum_{n=1}^{N} a_n \frac{< W_n|B_i >}{\|A_0\|\|B_i\|}, \tag{19}$$

then it selects the maximum $|s_i|$ by oscillating on the corresponding state $|B_i>$. Equation (19) can be rewritten more compactly as

$$T\hat{a} = \mu\hat{S}, \tag{20}$$

where the n-dimensional column vectors $\hat{a}$ and $\hat{S}$ have elements a_i and s_i, respectively; and the $N \times N$ matrix T has elements $T_{ij} = < W_i | B_j > / \|B_j\|$ and $\mu = \|A_0\| / \epsilon$. Hence, the components of the vectors $\hat{a}$ and $\hat{S}$ are, respectively, the expansion coefficients of the field $|A_0 >$ and the evaluated similarity parameters. As previously stated, the expansion coefficient a_n represents exactly the degree of presence in the input image I of the feature I_n. Therefore, to avoid errors during the selection of the stationary state, a direct proportionality between a_n and s_n has to be maintained. This amounts to requiring that $\hat{S} = k\hat{a}$, with k being an arbitrary real constant, so that (20) becomes

$$T\hat{a} = \lambda \hat{a} \qquad \text{with} \qquad \lambda = \mu k = k\|A_0\|/\epsilon \,, \tag{21}$$

which is a typical eigenvalue equation.

Equation (21) must hold for any value of the coefficients a_n of the vector $|A_0 >$. This means that the vector $\hat{a}$ must span the whole $\mathcal{R}^N$ space, i.e., any vector $\hat{a} \in \mathcal{R}^N$ must be an eigenstate of T. This can be obtained only if the matrix T is proportional to the identity matrix, precisely

$$T_{ij} = \frac{< W_i | B_j >}{\|B_j\|} = \lambda \delta_{ij} \,. \tag{22}$$

Therefore, a biorthogonality relation between stationary laser patterns $|B_i >$ and fields $|W_i >$ stored in the hologram H_2 has to be satisfied. This can be done by obtaining, first, mathematical expressions of fields W_i that are solutions of the matrix equations (22), then by using synthetic holography [Br1, Br2] for the physical realization of the hologram H_2. Equation (22) requires that the vectors $|B_i >$ be linearly independent and, in this case, can be satisfied in infinitely different ways. The orthogonality condition (22) is the element that characterizes the linear part of our system and distinguishes it, for example, from that of [LCT].

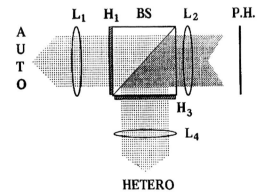

Fig. 5. The two holograms H_1 and H_3 allow for an auto-associative and a hetero-associative operation, respectively. BS = beam splitter, L_4 = lens.

Once the laser has approached a particular stationary state, the output of the laser is described by the field $|B_{\overline{n}}>$ of that state. In the backward direction, the stationary state $|B_{\overline{n}}>$ passes through the hologram H_2 and, due to the bi-orthogonality with fields $|W_i>$, only the pinhole corresponding to $|B_{\overline{n}}>$ will be bright. The pinhole constitutes the input for the correlator and the on-axis output is, therefore, the associated image $I_{\overline{n}}$ stored in H_1. The extracted image $I_{\overline{n}}$ represents exactly the feature that possesses the major degree of superposition (similarity) with the input I. This is a consequence of the bi-orthogonality relation (22) which, using (14), (18), and (17) and assuming for simplicity $\|I_n\| = 1$, implies the following linear relation between images and stationary laser patterns:

$$S(A_0, B_n) = cS(I, I_n), \tag{23}$$

where the constant c is defined as $\epsilon|\lambda|\|I\|/\|A_0\| = |k|\|I\|$. Hence, it is possible to recognize the presence of one particular image, previously stored as a feature in H_1, even if a distorted or partial version is present as input.

In addition to the auto-associative operation described before, as shown in Fig. 5, it is possible to obtain in parallel an ethero-association between the set $\{I_n\}$ and the content of a third hologram H_3.

Acknowledgments This research has been conducted within the framework of the ESPRIT BR Actions TOPP (Transverse Optical Patterns) and TONICS (Transverse Nonlinear Optics).

References

[Gdm] Goodman, J.W., Leonberger, F.J., Kung, S.Y., Athale, R.A.: Proc. IEEE **72** 1984, 850.
[Gbr] Gabor, D.: IBM J. Res. Dev. **13**, 1969, 156.
[VdL] Van der Lugt, A.B.: IEEE Trans. Inf. Theory, **IJ-10** 1964, 139.
[Hph] Hopfield, J.J.: Proc. Nat. Acad. Sc. USA, **79** 1982, 2554.
[Khn] Kohonen, T.: Self-Organization and Associative Memory. Springer-Verlag, 1984.
[CrG] Carpenter, G.A., Grossberg, S.: Appl. Opt. **26** 1987, 4919.
[Chn] Cohen, M.S.: Appl. Opt. **25** 1986, 2288.
[Ps1] Psaltis, D., Farhat, N.H.: Opt. Lett. **10** 1985, 98.
[Ps2] Abu-Mostafa, Y.S., Psaltis, D.: Scientific American, **88** March 1987.
[Sfr] Soffer, B.H., Dunning, C.J., Owechko, Y., Marom, E.: Opt. Lett. **11** 1986, 118.
[YrK] Yariv, A., Sze-Keung Kwong: Opt. Lett. **11** 1986, 186.
[ALF] Anderson, D.Z., Lininger, D.M., Feinberg, J.: Opt. Lett. **12** 1987, 123.
[And] Anderson, D.Z.: In AIP Conf. Proc. **n. 151** 1986, 12.
[LCT] Lalanne, P., Chavel, P., Taboury, J.: Appl. Opt. **28** 1989, 377.
[Hkn] Haken, H.: In Computational Systems – Natural and Artificial. Ed. Haken, H., Springer-Verlag, Berlin, 1987, 2.
[FuH] Fuchs, A., Haken, H.: In Neural and Synergetic Computers. Ed. Haken, H., Springer-Verlag, Berlin, 1988, 16.

[Vrn]   Vorontsov, M.A.: Proc. SPIE **402** 1991, 116.
[TmW]   Tamm C., Weiss, C.O.: J. Opt. Soc. Am. B **7** 1990, 1034.
[Br1]   Brambilla, M., Battipede, F., Lugiato, L.A., Penna, V., Prati, F., Tamm, C., Weiss, C.O.: Phys. Rev. A **43** 1991, 5090.
[Br2]   Brambilla, M., Lugiato, L.A., Penna, V., Prati, F., Tamm, C., Weiss, C.O.: Phys. Rev. A **43** 1991, 5114.
[BrL]   Brown, B.R., Lohmann, A.W.: IBM J. Res. & Dev. **13** 1969, 160.
[Lee]   Lee, H.W.: In Progress in Optics **XVI**. Ed. Wolf, E., North-Holland, Amsterdam, 1978, 119.
[Kgl]   Kogelnik, H.: In Lasers: A Series of Advances. Ed. Levine, A.K., **1**, Marcel Dekker, New York, 1966, 295.
[Lug]   Lugiato, L.A., Prati, F., Narducci, L.M., Oppo, J.L.: Opt. Comm. **69** 1989, 387.
[Br3]   Brambilla, M., Lugiato, L.A., Pinna, M.V., Prati, F., Pagani, P., Vanotti, P., Li, M. Y., Weiss, C.O.: Submitted for publication.
[An1]   Anderson, D.Z.: Opt. Lett. **11** 1986, 56.
[FrP]   Farhat, N.H., Psaltis, D.: In Optical Signal Processing. Ed. Horner, J.L., Academic Press, San Diego, 1987, Chapts. 2, 3.
[RBH]   Rajbenbach, H., Bann, S., Huignard, J.P.: Optical Computing Topical Meeting, Salt Lake City, March 1991.

7 Pattern and Vortex Dynamics in Photorefractive Oscillators

F.T. Arecchi[1,2], *S. Boccaletti*[1,2], *G. Giacomelli,* [1,3] *P.L. Ramazza*[1], *and S. Residori*[1]

[1] Istituto Nazionale di Ottica - Firenze - Italy
[2] Dept. of Physics - University of Firenze
[3] ITI "T. Buzzi" - Prato

Introduction

This paper reviews a recent series of investigations on pattern formation in an extended optical system. While early phenomena studied in lasers or other non-linear optical devices were either single mode or multimode, but with coherent phase relations among modes (mode locking), here we consider phenomena in which modes compete or mix in a weakly correlated or uncorrelated way, thus giving rise to "complex" patterns.

The core issue of any morphogenetic phenomenon, namely, how much of the pattern is due to the boundary condition and how much is intrinsic in the bulk properties, is here addressed and a comparison is made with pattern formation in chemistry and hydrodynamics.

After a decade of intensive exploration of low-dimensional chaos in nonextended, or extended but space uniform, dynamical systems [ArH], pattern formation phenomena and organization processes have become a topic of active interest in many scientific fields as diverse as materials science, hydrodynamics, chemistry, biology, and plasma physics [NcP, Hkn, Kr1, VdP, BGN]. In these cases, spatio-temporal structures arise by competition between local nonlinear dynamics and diffusive transport processes.

In 1952, Turing [Trn] established the theoretical possibility that stationary spatial patterns could organize in an initially homogeneous reaction-diffusion system. That pioneering work has generated many theoretical studies on the role of diffusion in the stability of steady states [NcP, ZhR, Vst, DlB]. Diffusive instabilities have also been proposed to account for propagating patterns in homogeneously oscillating systems [Kr1]. Near the oscillatory onset, the Ginzburg-Landau amplitude equation [LnG] can be reduced to the Kuramoto-Sivashinsky equation [Kr2, Svs], which describes slow space-time variations of the phase of the oscillator. Besides regular solutions, this equation exhibits chaotic solutions [PPP], a form of weak turbulence called phase turbulence [Kr1, Kr2]. More recently numerical simulation with the two-dimensional Complex Ginzburg-Landau equation (CGL) has provided an interpretation of weak turbulence in terms of the creation of spiral defects whose interactions and motion lead to complex spatio-temporal behavior termed "topological turbulence" [GGR, CGL, Grn, RcT, BPJ]. When more delocalized amplitude modes get excited, "defect-mediated turbulence" transforms into amplitude turbulence [Kr2, BLN].

Despite this great amount of theory, the experimental situation was not satisfactory. Only recently [Cst], evidence of Turing instability has been demonstrated in the laboratory. In fact, Turing instability to steady cellular patterns requires the diffusion coefficients of the different chemical species to be significantly different [KrO]. This occurs in activator-inhibitor competition, when the inhibitor diffuses faster than the activator, which is common to biological systems in which many processes are activated by enzymes immobilized in a matrix [Mnh], but is generally unrealistic for chemical reactions, such as the famous Belouzov-Zhabotinsky reaction [Bls], involving molecules in solutions with comparable diffusion coefficients.

As for propagative structures, they are common in excitable systems [MPH], however, they do not fit the Kuramoto model, because most of the chemical Hopf bifurcations are subcritical and, thus, lead to large amplitude relaxation oscillations [VdP]. Relatively few supercritical Hopf bifurcations have been observed [Arg]. Therefore, wave patterns observed in chemistry are not of the Kuramoto type, but they require an appropriate model introduced for extended excitable media [TsK].

In nonlinear optics, the competition arises between nonlinearities in the constitutive relation between local field and induced dipole and the "transport" effects of the field due to diffraction. If dissipation in the medium can be neglected, there is a purely conservative dynamics ruled by a nonlinear Schrödinger equation (NLS) whose solution provides coherent solitonic structures [Mln, FrW]. Accounting for unavoidable dissipation processes, NLS becomes a CGL, as will be shown in the forthcoming sections.

Pattern formation in optics then is a different process from pattern formation in chemistry. If the material medium is rather "broadband" both in frequency and in wave number, it simply provides a flat gain and patterns are shaped by the boundary conditions applied to the wave equation. Such is the case of the standard single-mode laser. The first step then is to study how the symmetry breaking at the boundary induces competition among different modes, limiting the model of the medium to the simplest nonlinearity that provides such a competition [Grn]. Because we are mainly interested in optical cavities fed by photorefractive crystals, which are narrow band both in frequency and wave number, we prefer to consider the detailed interaction from first principles.

This paper is organized as follows: Section 1, after general considerations of 1- and 2-dimensional pattern formation, describes the phenomenology of a photorefractive oscillator for different "aspect ratios." We use a term familiar in hydrodynamics to denote the ratio of transverse to longitudinal size of the optical cavity, which, as is well known, is represented by the Fresnel number

$$F = \frac{a^2}{\lambda L},\tag{1}$$

where a is the diameter of the pupil limiting the transverse size (x, y) of the cavity, L is the longitudinal size along the cavity axis z, and λ is the optical wavelength.

Section 2 introduces the phase singularities of a two-dimensional field. Because we consider cases in which the transverse variations are much smoother than the optical wavelength, we will see that in most experimental instances we can factor out the longitudinal dependence and write the field as

$$E(x, y, z) = E(x, y)e^{ikz}. \tag{2}$$

The two-dimensional field envelope can be written in complex notation at each point (x, y) of the complex plane as

$$E = \mid E \mid e^{i\Phi}. \tag{3}$$

Phase singularities are those points where the phase gradient has a nonzero circulation [Brr]:

$$\oint \nabla \Phi \cdot \mathbf{dl} \neq 0. \tag{4}$$

In Sect. 2, we describe how to measure these objects, what use to make of them, and how they are connected to pattern formation and competition. As a follow up, studying their statistics, we arrive at an important result, that is, the transition from boundary-dominated patterns to bulk-dominated patterns. These bulk-dominated patterns, just as Turing patterns, are ruled only by the medium property.

Section 3 presents the theoretical model and its numerical solutions.

This research has been partly supported by EEC contract, TONICS n^0 7118.

1 Pattern Formation and Complexity

1.1 The Multimode Optical Oscillator: 1-, 2-, 3-Dimensional Optics

Self-organization evolved from a qualitative concept [VnN] to a quantitative appraisal of a physical phenomenon as soon as physicists were able to pick up a low-dimensional laboratory system displaying a satisfactory agreement between its experimental behavior and the corresponding theoretical model. While theoretical models were worked out for nonequilibrium transitions in fluids and chemical reactions, the first clear experimental evidence was provided in the mid-sixties by the single-mode laser. This system displayed an outstanding agreement between theory and experimental tests conducted with photon-counting statistics [Ar1].

Later, similar threshold measurements were performed at the onset of instability phenomena in fluid convection or in chemical reactions, thus opening the field of the so-called "nonequilibrium phase transitions" [Slv]. However, in fluids it is straightforward to scale up the system size from small to large cells, thus making it possible to explore in many ways the passage from systems coherent (fully correlated) in space to systems made up of many uncorrelated, or weakly correlated, domains.

Crucial issues, such as (1) the passage from order to chaos within a single domain, and (2) possible synchronizations of time behaviors at different space domains, have been addressed in past years, with the general idea that space-time organization is what makes a large-scale object complex, that is, richer in information than the sum of the elementary constituents, due to an additional nontrivial cross information [And].

Thus far, such an investigation has not been possible in the optical field, because all coherent optics is based on the Schawlow-Townes original idea of a drastic mode selection [ScT]. The opposite, the thermodynamic limit of optics, was explored early in this century [VnL]. A confrontation of the two limits was shown experimentally in terms of photon statistics [Ar2] in 1965, however, what is in between has not been adequately explored, at variance with hydrodynamic instabilities or chemical reactions.

Because all coherent phenomena occur in a cavity mainly extended in a z-direction (e.g., the Fabry-Pérot cavity), we expand the field $e(r,t)$, which obeys the wave equation

$$\sqcap^2 e = -\mu \ddot{p} \tag{5}$$

(where $p(r,t)$ is the induced polarization), as

$$e(r,t) = E(x,y,z,t)e^{-i(\omega t - kz)}. \tag{6}$$

If the longitudinal variations are mainly accounted for by the plane wave, then we can take the envelope E as slowly varying in t and z with respect to the variation rates ω and k in the plane wave exponential. Furthermore, we call P the projection of p on the plane wave. By neglecting second-order envelope derivatives, it is easy to approximate the operator on E as

$$\sqcap^2 \rightarrow 2ik\left(\partial_z + \frac{1}{c}\partial_t\right) + \partial_x^2 + \partial_y^2, \tag{7}$$

as is usually done in the eikonal approximation of wave optics. This further suggests three relevant physical situations.

(1+0)-Dimensional Optics In this case, there is only a time dependence and no space derivatives, that is, $\sqcap^2 \rightarrow 2i\omega d_t$. Assuming that the laser cavity is a cylinder of length L, with two mirrors of radius a at the two ends, the cavity resonance spectrum is made of discrete lines separated by $c/2L$ in frequency (Fig. 1a), each line corresponding to an integer number of half-wavelengths contained in L, plus a crown of quasi-degenerate transverse modes at the same longitudinal wave number, with their propagation vectors separated from each other by a diffraction angle λ/a (Fig. 1b).

This case corresponds to a gain line narrower than the longitudinal frequency separation (so-called free spectral range) and to a Fresnel number of the order of

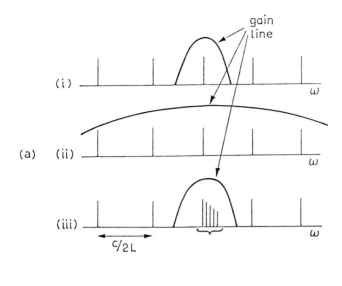

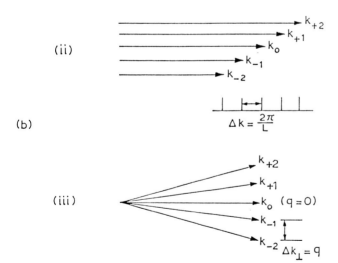

Fig. 1. (a) ω-space and k-space (b) pictures of the lasing modes in the (i) $(1 + 0)$, (ii) $(1 + 1)$, and (iii) $(1 + 2)$-dimensional cases.

unity, so that the first off-axis mode already escapes out of the mirror. Intuitively, F is the ratio between the geometric angle a/L of the view of one mirror from the other and the diffractive angle λ/a.

The resulting ODE, replacing the wave PDE, has to be coupled to the matter equations giving the evolution of P. In the simple case of a cavity mode resonant with the atomic line, we obtain the so-called Maxwell-Bloch equations [Ar3].

$$
\begin{aligned}
\dot{E} &= -\gamma_E E + gP, \\
\dot{P} &= -\gamma_\perp P + gEN, \\
\dot{N} &= \gamma_\| N - 2gEP + A,
\end{aligned}
\tag{8}
$$

where N is the population inversion (we have modeled the gain medium as a collection of two-level atoms), γ_E, $\gamma_\perp$, and $\gamma_\|$ are the loss rates of E, P, and N, respectively, g is the field-matter coupling constant, and A is the pump rate.

The equations in (8) are isomorphic to Lorenz equations for a model of convective fluid instability. These three, being nonlinear equations, provide the minimal conditions for deterministic chaos. However, time-scale considerations can rule out some of the three dynamical variables, yielding a dissipative dynamics with only one variable (fixed-point attractor) or two variables (limit-cycle attractor). Only when the three damping rates γ_E, $\gamma_\perp$, and $\gamma_\|$ are of the same order of magnitude, we have dynamics and, hence, the possibility of a chaotic motion (strange attractor). The above three cases have been classified as class A, B, and C lasers, respectively.

A comprehensive review of experiment and theory for these single-domain, (1+0)-dimensional systems is given in the book cited in [ArH], covering the period 1982-87, during which these space invariant instabilities were studied.

(1+1)-Dimensional Optics Here, we have a cavity thin enough to reject off-axis modes, but which is fed by a gain line wide enough to overlap many longitudinal modes. The superposition of many longitudinal modes means that one must retain the z gradient. Thus the wave equation reduces to

$$
(\partial_t + c\partial_z)E = GP,
\tag{9}
$$

where G is a scaled coupling constant.

Having a PDE, any mode expansion with reasonable wave number cutoff provides a large number of coupled $ODEs$, thus, it is immaterial whether P and N are adiabatically eliminated, as in class A and B lasers, or whether they keep their dynamic character as in class C lasers. Anyway, we have enough equations to see space-time chaos (STC).

(1+2)-Dimensional Optics As shown in Fig. 1 *iii*, let us consider a gain line allowing for a single longitudinal mode, but take a Fresnel number high enough to allow for many transverse modes.

We rescale the transverse coordinates x, y with respect to the cross-cavity size a, and the time t to the longitudinal photon lifetime $L/(cT)$, where T is the mirror transmittivity. The new variables are

$$x' = x/a \quad , \quad y' = y/a,$$
$$t' = \frac{t}{L/cT}. \tag{10}$$

Furthermore, we neglect the longitudinal gradient. Then, as shown by (7), the wave equation reduces to

$$\left(\partial_{t'} - i\alpha\nabla_\perp^2\right) E = GP, \tag{11}$$

where $\nabla_\perp^2$ is the transverse Laplacian and

$$\alpha = \frac{1}{4\pi FT}. \tag{12}$$

As in the $(1+1)$ case, (11) must be coupled with the material equation. If P has a fast relaxation toward a local equilibrium with the field, and if we expand its dependence to the lowest orders, we have a relation as

$$P = aE - b\mid E \mid^2 E. \tag{13}$$

Introducing this into (11), one has a nonlinear Schrödinger equation (NLS), which has been recently considered in many theoretical investigations [Mln, FrW].

On the other hand, important considerations have been developed for the complex Ginzburg-Landau equation (CGL). This can be written as

$$\partial_t u = (\alpha_1 + i\alpha_2)\nabla^2 u - \mu u - (1 - i\beta)\mid u \mid^2 u. \tag{14}$$

The CGL includes NLS (for $\alpha_1 = 0$), the chemical reaction-diffusion equation (for $\alpha_2 = 0$), and the single-mode laser equation (for $\alpha_1 = \alpha_2 = 0$).

Writing the complex field as $u = \mid u \mid e^{i\phi}$, under broad assumptions one may localize regimes where the modulus has small variations, whereas the phase has crucial changes described by the Kuramoto-Sivashinsky equation [Drn].

The CGL and the corresponding phase equation have played a fundamental role in understanding many features of hydrodynamic turbulence [CFT], including the formation of cellular patterns [Shr] and STC [HhS].

1.2 The Photorefractive Ring Oscillator. How to Control the Fresnel Number

In early quantum optical investigations, coherent optical oscillators have been considered as discrete physical systems in which only one mode, or a few at most, can survive [Lmb]. Even though photon statistics provide an accurate tool to test the noise dependence of the laser around the threshold of oscillation [Ar1], and even though a variety of instabilities and routes to optical chaos have been described [ArH], for many years optical devices have defeated the search for spatial dependence.

We show here how this can be achieved [Ar4]. We seed a ring cavity with a photorefractive gain medium pumped by an argon laser to study the temporal and spatial features of the generated field. By varying the size of a cavity pupil, it is possible to control the number of transverse modes that can oscillate. The experimental setup, shown in Fig. 2, consists of a ring cavity with photorefractive gain. The gain medium is a 5 x 5 x 10 mm BSO (bismuth silicon oxide) crystal to which a direct current (dc) electric field is applied. The crystal is pumped by a cw argon laser with an intensity around 1 mW/cm^2. The basic cavity configuration consists of four high-reflectivity dielectric mirrors and a lens L' of 500 mm focal length, which enhances the cavity mode stability, providing a near-confocal configuration.

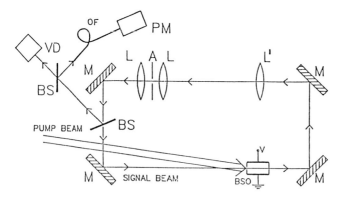

Fig. 2. Experimental setup: Video camera (VD) records the wavefront patterns; photomultiplier (PM) measures the time evolution at a point selected by fiber OF.

The Fresnel number of the cavity is controlled by a variable aperture. A pinhole of 300 μm diameter is inserted in the optical path between two confocal lenses L of short focal length (50 mm). Small displacements of the pinhole along the optical axis yield a continuous change of the aperture to spot size ratio, and consequently, inhibit a different number of transverse modes. The effective Fresnel number F is the ratio of the area of the diffracting aperture that limits the system (pupil) to the size of the fundamental Gaussian-mode spot, evaluated in the plane where the aperture is placed. F can be varied in the range from 0 to approximately 100 (Fig. 3). This corresponds roughly to the variation of the number of transverse modes that can oscillate.

The fundamental mode spot size $w_0(z)$ is evaluated in each plane along the propagation direction z, using the matrix method for the propagation of Gaussian beams. The Fresnel number F is the ratio between the areas of the pinhole and that of the fundamental mode spot in the same plane

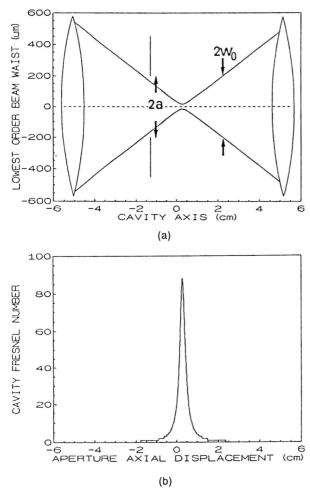

Fig. 3. (a) Spot size of the fundamental TEM_{00}, resulting from the matrix method calculation between the two confocal lenses. (b) Effect of the variable pupil on the cavity Fresnel number F. The horizontal axis reports the displacements of aperture A away from the focus of the lens. The slight asymmetry is due to the presence of lens L'.

$$F = \frac{a^2}{w_0^2(z)}. \tag{15}$$

Notice that in a standard confocal cavity, if the pinhole is positioned at the minimum beam waist $w_0 = \sqrt{\lambda l/2\pi}$, then (15) coincides with (1), besides a factor of the order of unity.

Because the overall size of the Gaussian mode of order n is $w_n \simeq \sqrt{n}w_0$, the highest allowed mode (of order $\bar{n} = n_{max}$) would have a size $w_n = a \simeq \sqrt{\bar{n}}w_0$, so that

$$F \simeq \frac{\bar{n}w_0^2}{w_0^2} = \bar{n}. \tag{16}$$

Thus the Fresnel number F gives the maximum order of the transverse modes that can oscillate. F can be varied in the range from 0 to approximately 100. Correspondingly, the total number of transverse modes allowed by diffraction scales is F^2.

For the n order mode, the spatial separation between its zeros is of the order of

$$\delta = 2\frac{w_n}{n} = 2\frac{\sqrt{n}w_0}{n} = 2\frac{w_0}{\sqrt{n}}. \tag{17}$$

1.3 Periodic (PA) and Chaotic (CA) Alternation and Space-Time Chaos

Transverse intensity patterns corresponding to increasing Fresnel numbers F are shown in Fig. 4, together with the spatial autocorrelation functions of the intensity fluctuations around the local average.

The low F limit ($F \leq 4$) corresponds to a time alternation between pure cavity modes (Fig. 4a), yielding a spatial correlation length of the order of the transverse size D of the beam. For high F ($F \simeq 15$), on the contrary, the signal appears as a speckle-like pattern irregularly evolving in space and time (Fig. 4c), with a short correlation length ($\xi/D < 0.1$). The transition between these two limits is characterized by a continuous variation of the ratio ξ/D. A generic intermediate situation is shown in Fig. 4b. An example of alternating pure mode configurations is given in Fig. 5.

To study the time behavior of the system, we perform a local measurement of intensity versus time by placing an optical fiber in an arbitrary point on the wavefront. Figure 6a shows the results of this measurement for $F = 7$; for this value of F there is one cavity mode oscillating at a time, so that each level is encoded in a local intensity level. Identifying each mode with its azimuthal quantum number (defined as half the number of nodes along a circumference), we have here modes 7, 6, 5, 4, 3, 2, 1, and 0 alternating on a time scale of seconds, that is, of the order of the dielectric relaxation time of BSO. In Fig. 6a, the alternation is irregular (CA). To control the time sequence, making it regular (PA), we limit the number of interacting modes by replacing the circular diaphragm with an annular one. This, in fact, can be done by inserting an axial stop. The corresponding result is shown in Fig. 6b. The periodic alternation between modes is a behavior peculiar to low Fresnel numbers. It is observed down to the minimum F for which the cavity can oscillate.

Let us summarize our findings. The cylindrical geometry of the cavity constrains the symmetry of the output field. The pumping process, however, breaks the $O(2)$ cavity symmetry, introducing a privileged plane defined by the propagation vectors of pump and signal fields. The oscillator yields field patterns

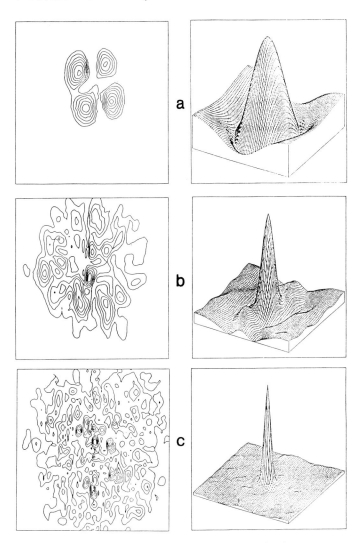

Fig. 4. Intensity distribution of the wavefront (left) and spatial autocorrelation function (right) for increasing Fresnel number. (a) $F \simeq 1$, one single mode at a time is present, ratio between coherence length ξ and frame size D is $\xi/D \simeq 1$; (b) $F = 7$, $\xi/D \simeq 0.2$; (c) $F = 10$, $\xi/D \simeq 0.06$.

varying in time. By changing the size of the cavity pupil, two different dynamical regimes are observed. For large pupils, the field displays a complex pattern, which may be expanded in a large number of solutions of the free propagation problem (the so-called cavity modes). For small pupils, the field at any time is made of a single mode, however, a small number of modes (from two to about ten) can alternate in time. Thus, the alternation phenomenon consists of an ordered sequence of quasi-stationary modes. Depending on some control param-

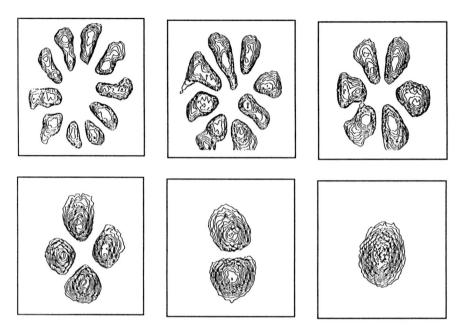

Fig. 5. Intensity patterns of the pure modes in their order of consecutive appearance in a cycle of periodic alternation at $F = 5$.

eter, the persistence time of each mode is either regular (periodic alternation) or irregular (chaotic alternation). Away from the narrow switching time intervals, the amount of mode mixing is negligible.

A phenomenon similar to CA, called chaotic itinerancy, was introduced by Ikeda and colleagues [IMO], Otsuka [Ots], Kaneko [Knk], and Tsuda [Tsd] in dealing with numerical solutions of different classes of model equations, namely, a one-dimensional laser [IMO], an array of coupled lasers [Ots], globally coupled iteration maps [Knk], and nonequilibrium neural networks [Tsd]. In fact, the latter phenomenon includes erratic jumps among the available quasi-stationary states, whereas CA keeps the sequence ordering.

Increasing the value of the control parameter, we enter a new regime, called spatio-temporal chaos, where a large number of modes coexist. This regime has been characterized on very general grounds by Hohenberg and Shraiman [HhS]. Suppose that we have a generic field $u(r, t)$ ruled by a PDE, including nonlinear and gradient terms. Such is indeed the situation of our (1+2)-dimensional optical system. Let us take the field of deviations away from the local time average

$$\delta u(r, t) = u(r, t) - < u(r, t) >, \tag{18}$$

where $< ... >$ denotes time average. Under very broad assumptions, we can take the leading part of the correlation function as an exponential, that is,

$$C(r, r') = < \delta u(r, t)\delta u(r', t) > \simeq e^{-|r-r'|/\xi}. \tag{19}$$

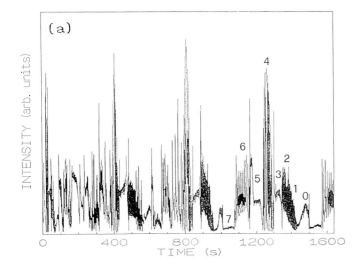

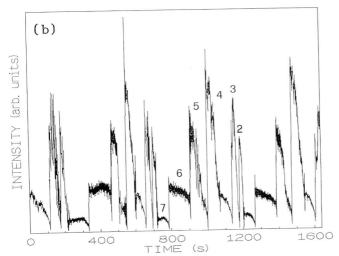

Fig. 6. Time records of local intensity (samples collected at 10 Hz rate) at $F = 5$. (a) with the low-pass filter (CA), (b) with the bandpass filter (PA).

Whenever the correlation length ξ is larger than the system size $L(\xi > L)$, we have low-dimensional chaos, that is, even though the system can be chaotic in time, it is coherent in space (single mode, in a suitable mode expansion). The corresponding chaotic attractor is low dimensional. In the opposite limit of $\xi << L$, a local chaotic signal is not confined in a low-dimensional space. However, a new outstanding feature appears. If we collect a local time series of data $\delta u(r, t)$ at a given point r, the corresponding statistical distribution $P(\delta u(r, t))$ is strongly non-Gaussian. There is no wonder about that, after all, δu stems from a strongly nonlinear dynamics. But if we now Fourier transform $\delta u(r, t)$, the corresponding dynamical variable $\delta u(q,t)$ in wave-number space displays Gaussian statistics. This is somewhat surprising, because a Fourier transform is a linear operation and cannot introduce a Gaussian property when that property was absent.

Considering, however, the Fourier transform

$$\delta u(q,t) = \int dr e^{-iqr} \delta u(r,t), \tag{20}$$

we realize that up to a cut-off wave number $q_c = 1/\xi$, the phase factor qr changes very little if r is confined within a segment smaller than ξ. Thus, in the phased sum (20), we can replace δu with its coarse-grained approximation

$$\delta \bar{u}(r,t) \equiv \frac{1}{\xi} \int_r^{r+\xi} \delta u(r',t) dr', \tag{21}$$

where $\delta \bar{u}$ is an average over a correlation length.

With this replacement (20) becomes a sum of uncorrelated objects, and hence, it has a Gaussian distribution by the central limit theorem. We can say that the coarse-graining operation has been the nonlinear device distorting the statistics.

To prove such a conjecture, we have measured (Fig. 7) the normalized skewness $M_3/M_2^{3/2}$ and the flatness $M_4/3M_2^2$ for local and spectral intensity fluctuations (M_i is the ith-order moment of the statistical distributions). It is evident that the local flatness (LF) and skewness (LS) have deviations from the Gaussian values (1 and 0, respectively), which are much larger than the residual fluctuations. On the contrary, in k space both spectral flatness (SF) and skewness (SS) are centered at 1 and 0, respectively, showing evidence of STC.

Now we can see the role of complexity as mutual information, versus an incoherent set of simple objects. In our case, a non-Gaussian set of spots gives an overall complex pattern, because non-Gaussianity means mutual conditions. On the contrary, in the case of a speckle field, the Gaussian distribution [Ar2] displays the randomness character of the speckles, which are "simple" as a random number or a Boltzmann gas.

2 Phase Singularities, Topological Defects, and Turbulence

2.1 Phase Singularities in Linear Waves. Speckle Experiments

As stated in the Introduction, a phase singularity is a point around which the circulation of the phase gradient is a multiple of $\pm 2\pi$. We call topological charge the multiplicity number. In the case of the wave equation, only ± 1 charges are stable [Brr]. Let me quote from Berry [Brr]:

"Singularities, when considered in the modern way as geometric rather than algebraic structures, are *morphologies*, that is form rather than matter; and waves are morphologies, too (it is not matter, but form that moves with a wave). Therefore, singularities of waves represent a double abstraction – forms of forms,

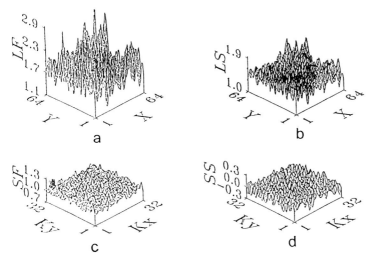

Fig. 7. (a) and (b) Normalized local flatness (LF) and skewness (LS) for intensity fluctuations in (x, y) space. (c) and (d) Normalized spectral flatness (SF) and skewness (SS) for intensity fluctuations in (k_x, k_y) space.

as it were – so, it comes as something of a surprise to learn that they represent observable phenomena in a very direct way."

The most important features of wavefronts are their singularities, which correspond to singularities of the phase function $\Phi(\mathbf{r}, t)$. The nature of these singularities is determined by the fact that E is a smooth single-valued function of its variables. Single valuedness implies that during a circuit C in space-time Φ may change by $2m\pi$, where m is an integer. Suppose m is not zero, and let C be shrunk to a very small loop in such a way that m does not change. Then C encloses a singularity, because Φ is varying infinitely fast. The smoothness of E now implies that this can happen only where $E = 0$, i.e., where Φ (3) is indeterminate. Because the vanishing of E requires two conditions ($\mathrm{Re}E = \mathrm{Im}E = 0$), these phase singularities are lines in space or points in the plane (Fig. 8a). Sometimes we call the phase singularities "wavefront dislocations" or, by analogy with fluid dynamics, vortices.

Let us consider a random complex field $E(\mathbf{r}$, where $\mathbf{r} = (x, y))$. If it is formed by the interference of a large number of independent components, $\mathrm{Re}E$ and $\mathrm{Im}E$ are two independent random functions with Gaussian statistics [Ar2].

The zeros of the function $\mathrm{Re}E(x, y)$ determine a number of curves in the (x, y) plane (see Fig. 8b). There is another set of curves corresponding to $\mathrm{Im}E(x, y) = 0$, and now the intersections of curves of one family with those of the other give discrete points in which $\mid E(x, y) \mid= 0$. If we consider the problem of the propagation of such zeros along the direction z in accordance with the wave equation, the discrete points in the (x, y) plane are converted into lines. It is clear that in general these lines do not intersect in three-dimensional space. Moreover, a given line can not appear singly at the same plane $z = \mathrm{const}$, nor

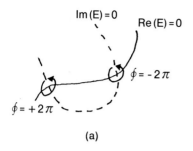

(a)

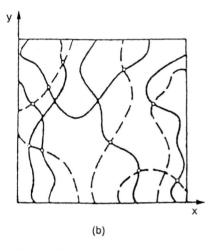

(b)

Fig. 8. (a) Pair of phase singularities or wavefront dislocations of opposite topological charge. (b) System of dislocations in a random field. Solid lines $\mathrm{Re}(E) = 0$ and dashed lines $\mathrm{Im(E)} = 0$.

can it disappear singly. Zeros in the pattern must appear or be annihilated in pairs.

Up to now, we spoke of a random speckle pattern. It is clear, however, that the topological dimensional arguments do not depend on the nature of the interfering fields. The only difference is that for regular fields, zeros may not occur at all.

The difference $N_+ - N_-$ of the numbers of zeros with positive and negative charges is conserved in the process of propagation. On the average, in a cross section of a speckle field $N_+ = N_-$, because the beam is statistically homogeneous.

We now derive the general expression for the mean number of zeros per unit area, $N = N_+ + N_-$, for a statistically homogeneous random process [BrZ]. The total number of zeros in an area S of the x, y plane can be written as.

$$NS = < \int dxdy \delta(E_1(x,y))\delta(E_2(x,y)) \mid \partial(E_1,E_2)/\partial(x,y) \mid > . \qquad (22)$$

Here $E = E_1 + iE_2$. The angle brackets denote averaging over an ensemble of random fields. Notice that each zero point of the total field will give (before averaging) a contribution unity to the right side of the equation. Incidentally, positive and negative zeros correspond, respectively, to positive and negative signs of the Jacobian $G = \partial(E_1,E_2)/\partial(x,y)$. The ensemble average can be rewritten as (denoting the gradients as $\partial E_1/\partial x = E_{1x}$, etc.)

$$
\begin{aligned}
N &= \frac{1}{S} \int dxdy \int dE_1 dE_2 dE_{1x} dE_{2x} dE_{1y} dE_{2y} \\
&\quad \cdot W_6(E_1, E_2, E_{1x}, E_{1y}, E_{2x}, E_{2y})\delta(E_1)\delta(E_2) \mid G \mid \\
&= \int dE_{1x} \ ... \ dE_{2y} W_6(0, 0, E_{1x}, E_{1y}, E_{2x}, E_{2y}) \mid G \mid .
\end{aligned}
\qquad (23)
$$

Here W_6 is the joint probability of the quantities E_1, E_2, and their gradients at the given point. Making the Gaussian assumption, W_6 is factored out in terms of correlations of the complex field. By the Van Cittert-Zernicke theorem [BrW], the correlation is

$$< E^*(r_1)E(r_2) >= I \int j(\theta) e^{ik\theta(r_2-r_1)} d^2\theta, \qquad (24)$$

where $j(\theta)$ is the normalized angular spectrum and $\theta = (\theta_x, \theta_y)$. Furthermore,

$$< E^*(r_1)\frac{\partial E(r_2)}{\partial x_i} >|_{r_1=r_2} = ik < \theta_i >= ik \int j(\theta)\theta_i d^2\theta. \qquad (25)$$

By a suitable rotation of the z axis, we can make $< \theta_x >=< \theta_y >= 0$, that is, choose the new z axis in the direction of the center of gravity of the angular distribution. Then, the complex gradients $\partial E/\partial x$ and $\partial E/\partial y$ are independent of the field $E(r)$ itself at each point (x,y). By a rotation of axes in the x,y plane, the correlation matrix

$$< \theta_i\theta_k >= \frac{1}{k^2} < \frac{\partial E^*(r_1)}{\partial x_i}\frac{\partial E(r_2)}{\partial x_k} >|_{r_1=r_2} = \int j(\theta)\theta_i\theta_k d^2\theta \qquad (26)$$

can be transformed to principal axes, hence all three complex quantities $E, \partial E/\partial x$ and $\partial E/\partial y$, are mutually independent at the given point. In this system of coordinates, with $< \theta >= 0$ and matrix $< \theta_i\theta_k >$ diagonal, W_6 becomes

$$W_6(E_1, E_2, E_{1x}, E_{1y}, E_{2x}, E_{2y}) =$$

$$\frac{1}{\pi^3 k^4 < \theta_x^2 >< \theta_y^2 > I^3} e^{-\frac{(E_{1x}^2+E_{2x}^2)}{k^2 <\theta_x^2> I}} e^{-\frac{(E_{1y}^2+E_{2y}^2)}{k^2 <\theta_y^2> I}} e^{-\frac{(E_1^2+E_2^2)}{I}}. \qquad (27)$$

Averaging, we find

$$N = \frac{k^2}{2\pi} (< \theta_x^2 >< \theta_y^2 >)^{1/2}. \tag{28}$$

Because the correlation radius of the speckle field, i.e., the transverse speckle size, is $\sim \lambda/\Delta\theta$, where $\Delta\theta$ is the angular divergence of the beam, the previous relation shows that the dislocation density coincides with the number of speckles per unit area. This was verified in a series of experiments [BrZ], in which the speckle field was obtained by transmitting a laser beam through a distorting phase plate. The structure of the speckle field wavefront was investigated by interfering with a plane reference wave directed at a certain angle. The fringe separation corresponds to the tilting angle between speckle and reference fields. The bending of the fringes corresponds to the curvature of the wavefront under investigation, while termination of a fringe or birth of a new fringe corresponds to a phase singularity (positive or negative charge, respectively). The density N scales like a^2 (a = diaphragm diameter on the phase plate), that is, linearly in F, thus the total number of dislocations Na^2 scales like F^2.

2.2 Phase Singularities in Nonlinear Optics: Scaling Laws

In the previous section, we introduced phase singularities in linear wave fields and discussed their geometrical properties. In nonlinear physics, singular points can have the dynamical role of topological defects that mediate the transition between two different types of symmetry.

Any symmetry breaking is accompanied by the appearance of a defect. If the defect is a space structure, as grain boundaries or point defects in crystals or convective structures, then the defect is called a "structural defect." In wave patterns there is no strict equilibrium analogue, because they appear in space-time. In such a case, a break of the phase symmetry is called a "topological defect" [Cll]. However, we should not call any singularity a defect, only those localized at the edge between two patterns with different symmetries.

Let us illustrate this idea with reference to a recent experiment on thermal convection in a fluid [Clb]. A tiny change of the temperature gradient in a Rayleigh-Benard cell filled with water induces a competition between rolls and hexagons. Hexagons can be seen as the superposition of three sets of rolls at 120° from each other. If the amplitudes of the three sets are equal (Fig. 9a), regular hexagons are formed. If at a point P, two roll amplitudes go to zero, from that point on only the third set of rolls is present, thus we have a transition from hexagons to rolls. In P (Fig. 9b), the disappearance of two rolls is signed locally by a pentagon-heptagon pair. But, if we decompose the hexagonal pattern into the sum of the three roll amplitudes

$$T(x,y) = \sum_{j=1}^{3} A_j(x,y)e^{(i\mathbf{K}_j \cdot \mathbf{x})} + c.c \tag{29}$$

with $A_i = | A_i | e^{i\phi_i}$, then in P, where $| A_1 | = | A_2 | = 0$, $A_3 \neq 0$, the circulations of ϕ_1 and ϕ_2 have jumps of $\pm 2\pi$, respectively, whereas ϕ_3 has no singularity.

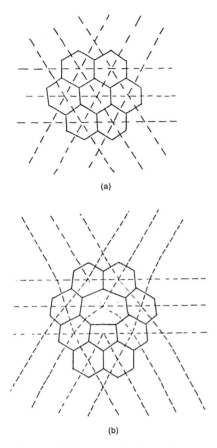

(a)

(b)

Fig. 9. (a) System of rolls giving rise to hexagons. Dashed lines are loci of minima. When the three roll systems have equal amplitude, they give rise to hexagons. (b) When $\mid A_1 \mid = \mid A_2 \mid = 0$ (thus Φ_1 and Φ_2 singular) a pair penta-heptagon appears. The core of the defect is on the side common to pentagon and heptagon, which in fact is aligned along the surviving roll system.

Thus, we see that a defect implies a phase singularity, but not vice versa. Notice, however, that the singularity appears in the filtered measurement of only one component (either 1 or 2) [Clb], whereas in optics, heterodyne provides the phase of the total field.

In our case [Ar5], as we will report later in this section, for low Fresnel numbers, we see a transition in time mediated by a phase singularity. Presumably in generalized space-time the difference between structural and topological defects disappears.

The role of defects in mediating turbulence in hydrodynamic systems with large aspect ratios has been investigated in fluid thermal convection [Clb, AhB], nematic liquid crystals [RbJ], surface waves [GlR], analytic treatments [Kws], and numerical simulations [CGL] of partial differential equations in 2+1 space-time dimensions. The possible role of defects in mediating turbulence in nonlinear optics has been discussed theoretically [CGR].

In nonlinear optics, we have recently shown experimental evidence of phase singularities [Ar5]. Their positions and the scaling of their separations, number, and charges with the Fresnel number allow a classification of patterns (see Sect. 2.4). However, we have not yet investigated their role in pattern competition at high Fresnel numbers, where patterns of different symmetries are simultaneously present. For this reason, we prudently speak of phase singularities or wavefront dislocations, avoiding for the time being, calling them topological defects .

At variance with the material waves, which are easily visualized in terms of matter displacements, in the case of an optical field a phase measurement requires heterodyning against an external reference. Phase information is extracted by beaming the signal with a reference beam onto a CCD video camera.

By a suitable algorithm [TIK], we reconstruct the instantaneous surfaces of phase as shown in Fig. 10, where the phase surface of a doughnut mode is a helix of pitch 2π around the core (vortex).

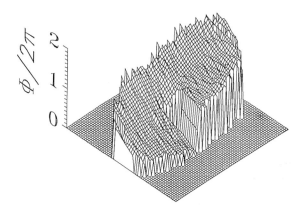

Fig. 10. Reconstruction of the instantaneous phase surface, perspective plot for a doughnut mode.

When more than one vortex is present, to resolve and count each vortex, we tilt the reference beam so that the video signal is now given by

$$I(x,y) = A^2 + B^2 + 2AB\cos(Kx + \Phi(x,y)), \tag{30}$$

where A and B are the amplitudes of reference and signal field, K is the fringe frequency due to tilting, x is the coordinate orthogonal to the fringes and Φ is the local phase. This way, a phase singularity appears as a dislocation, and the topological charge is visually evaluated.

Figure 11a shows an intensity pattern as a doughnut with a phase singularity at the center. On a long time scale (several seconds), the doughnut splits into two separate intensity maxima with no phase singularity (Fig. 11b), and later it reappears, but the singularity now has the opposite charge (Fig. 11c). In such a case, we can speak of a transition from an azimuthal wave traveling clockwise to traveling counterclockwise, with a standing wave (equal amplitudes of the two traveling waves) in between. According to the formal definition of defect, we do have a defect here as a sharp space variation in the amplitudes of the two competing traveling waves, but this occurs in time, whereas in space, the singularities of Figs. 8a and 8c do not delimit different symmetries. Strictly speaking, whether this is or is not a defect, seems to be a matter of terminology.

There are profound differences that make the phenomena observed in nonlinear optics qualitatively different from the linear dislocations of Sect. 2.1. The main ones are as follows:

(1) Our dislocations are closely linked to the coherent photorefractive field. This is a threshold phenomenon, which disappears for an applied dc field below 5 kV/cm.

(2) In the course of time, the nonlinear dislocation pattern evolves, whereas the dislocation pattern of a speckle field stands still, and it changes only by modification of the scattering medium (e.g., rotation of the phase plate).

(3) The linear dislocation pattern depends only upon the random superposition of the scattered wavelets [Brr], and indeed it is absent in the near field. In our experiment, instead of the dislocation pattern's dependence on the mode, dynamics do not require a Rayleigh length to be observed.

We digitize the fringe system and count those defects separated by at least one fringe, in the region where fringes can be resolved. Fig. 12a shows a configuration with an overall unbalance in the topological charge. A model of phase singularities in optics [Brm] displays regular patterns of defects with total nonzero charge. A heuristic explanation of Fig. 12a is that for small F, the dynamics are strongly boundary dependent. Consequently, we conjecture that an increase of F should eventually yield the thermodynamic limit of paired charges. This is indeed the case as shown in Fig. 12b, which refers to a high F, and where the charge unbalance $U = |n_+ - n_-|$ ($n_\pm$ – numbers of charges of the two signs) has become very small compared to the total number $N = n_+ + n_-$ of charges.

By averaging over a large number of frames for each F, we report in Fig. 13 a and b the mean number $< N >$ of singularities per frame and the mean nearest neighbor distance $< D_1 >$ versus F. $< N >$ and $< D_1 >$ have a power law dependence on F with exponents close to 2 and -0.5. For convenience, we also plot the second and a third neighbor separations $< D_2 >$ and $< D_3 >$.

Fig. 13c gives the excess U normalized to the total singularity number N (U_n). For small apertures, the dynamics are strongly boundary dependent and the excess is large. For increasing F, U_n decreases as a power law with exponent close to -1.5. Furthermore, measuring the space correlation functions of the intensity fluctuations, the corresponding correlation length ξ scales also as a power of F with an exponent close to -0.5 (Fig. 13d).

The spatial disordering of singularity positions is associated with the passage to STC. Viewing the dynamics of the optical field in the STC regime as ruled by a two-dimensional fluid of interacting defects [CGL], we expect that each one

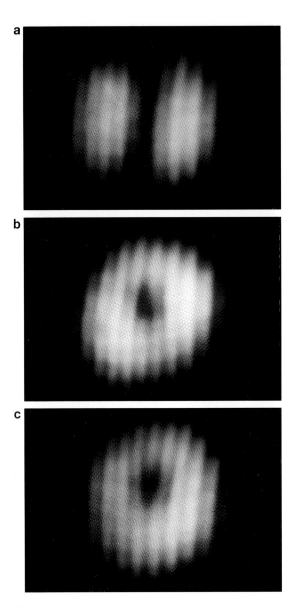

Fig. 11. Temporal sequence showing a mode switch with a vortex visualized as a dislocation via a tilted reference beam. Inversion of topological charge from (a) to (c).

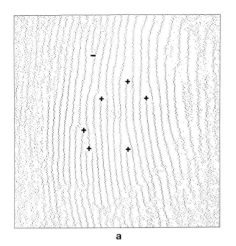 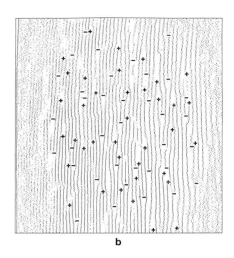

a   b

Fig. 12. Two examples of experimental configurations obtained by digitizing the fringe maxima are (a) $F \simeq 3$, six defects of equal topological charge against one of opposite charge; (b) $F \simeq 10$, about 70 defects of opposite charges, with residual small charge unbalance.

occupies an area of diameter D. Because a phase singularity must be associated with a zero crossing of real and imaginary parts of the field, it follows that all intensity zeros are singularities. But the diffractive treatment of optical cavities shows that the number N of intensity zeros for the highest allowed mode scales as the square of F ($N \sim F^2$) [Sgm]. On the other hand, if a is the pupil aperture of the optical system, and D the average inter-defect separation, we expect $N \sim a^2/D^2$, and, because $a^2 \sim F$, then $D \sim F^{-0.5}$. Such scaling laws are approximately verified in Fig. 13; however, there are sizable deviations between heuristic and experimental exponents. A qualitative explanation of the first deviation is that $N \sim F^2$ holds only for the highest mode allowed by F, but instead our dynamics in the STC regime imply a strong configuration mixing. The $\xi \sim F^{-0.5}$ dependence of Fig. 13d shows that indeed STC is closely linked to the singularity dynamics. We also can justify the F dependence of U_n. Assume that unpaired defects are mainly created at the boundary, while in the bulk, pairs with compensated charge are created and destroyed. We conjecture that the total number N_c of boundary defects in the parametrial region of area $a.D$ scales as $N_c \sim a.D/D^2 \sim a/ < D_1 > \sim F^1$, and the corresponding unbalance is $U \sim \sqrt{N_c} \sim F^{0.5}$. Hence, the normalized unbalance scales as $U_n \sim F^{0.5-2} = F^{-1.5}$, and is in good accord with the experiment.

   F scalings of the dislocation numbers and their mean separation roughly equivalent to those reported in Fig. 13 are also found for linear dislocations [BrZ]. The most crucial test of the nonlinear nature of the dislocation reported is given by its time dependence. We select a small box of side ξ (a correlation domain), where generally there is zero or one defect present, and measure the occurrence

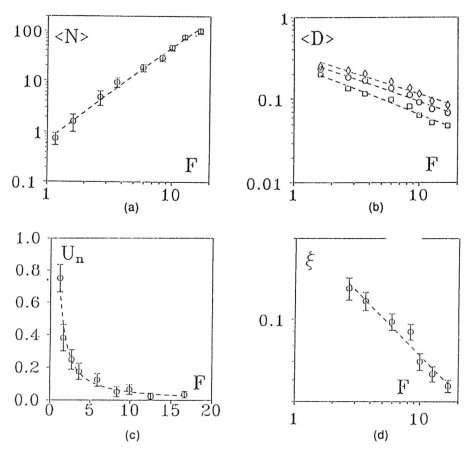

Fig. 13. (a) Mean number of defects $< N >$ versus F. Power law with exponent 2. (b) Mean neighbor separations $< D_1 >$ (squares), $< D_2 >$ (circles), and $< D_3 >$ (diamonds) versus F. Power law exponents close to -0.5. Here and in the next figures all distances are normalized to the maximum size of the acquisition frame (512 x 512 points), and error bars correspond to the data spread over many runs. (c) Normalized charge unbalance $U_n = |n_+ - n_-|/(n_+ + n_-)$ versus F. n_+ (resp. n_-) is the total number of positive/negative defects. Exponent close to -1.5. (d) correlation length of the intensity fluctuations ξ versus F. Exponent close to -0.5.

time of events, where an event is the entrance of a defect into the box. This way, we build a sequence of time intervals, each defined by two successive events.

The corresponding mean separation $< T >$ versus F is plotted in Fig. 14a for a fixed pump intensity P. Because for any setting of the control parameters F and P, the time $< T >$ is of the same order as the long time scale that characterizes the mode competition, we infer that mode jumping is mediated by the defect dynamics, as expected from the theory [CGL]. Fig. 14b shows a linear dependence of $1/ < T >$ on the pump intensity. This effect, together with the threshold dependence on the dc field, are clear evidence of the nonlinear nature of the defects.

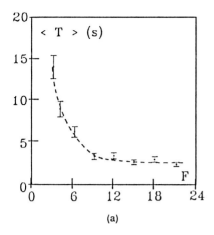

(a)

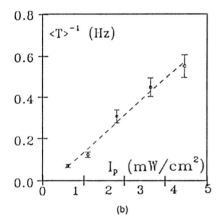

(b)

Fig. 14. (a) Mean separation time $< T >$ between defect occurrence within a correlation domain versus F, at a fixed pump intensity $P = 2.25$ mW/cm^2. (b) Mean frequency of occurrence $1/ < T >$ versus P, at a fixed Fresnel number $F = 8$.

2.3 Comparison of Vortex Statistics in Speckle and Photorefractive Patterns

As a statistical indicator, we study the fluctuations of the total number of vortices in the nonlinear field. For this purpose a set of 1024 frames of 512 x 512 pixels, separated in time by 5 sec, is acquired with the video camera. The sampling rate is much longer than the time scale of the signal, which in these conditions is of the order of 0.5 sec [Ar5, Ar4]. This ensures the statistical independence of the data. Furthermore, to avoid boundary effects, we choose an area of observation of 128 x 128 pixels at the center of the frame. Counting the number of vortices within this area, we obtain the histogram shown in Fig. 15 a.

A set of 1024 frames is acquired in the same way for the linear experiment. The diffuser translates at a constant velocity, and a time interval of 5 sec between

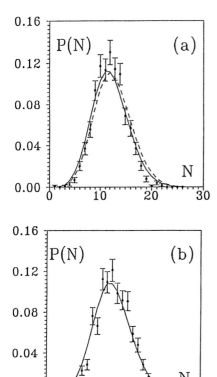

Fig. 15. Probability distribution for the number of defects; points are experimental data for (a) the nonlinear case and (b) the linear case. Solid lines are Poissonian fits. In (a) the Poissonian is compared with the theoretical distribution resulting from (32) (dashed curve).

two successive frames is sufficient to ensure the complete decorrelation of the frames. The number of vortices is counted in an area of 128 x 128 pixels, as for the nonlinear case. The resulting histogram is shown in Fig. 15b.

In Fig. 15a the continuous curve represents the result of a theoretical prediction by Gil and colleagues [GLM]. Their hypothesis is that vortices can only be created and annihilated by pairs, with rates of creation $\Gamma_+ = \alpha$ and annihilation $\Gamma_- = \beta n^2$, where n is the number of pairs present. This leads to the distribution

$$P(n) = \gamma \frac{\bar{n}^{2n} e^{-2\bar{n}}}{(n!)^2}, \tag{31}$$

where $\bar{n}$ is the mean number and γ is a normalization constant. Equation (31) is a square Poissonian in the number n of pairs. When reported to the number $N = 2n$ of vortices, it becomes

$$P(N) = \gamma \frac{(\bar{N}/2)^N e^{-\bar{N}}}{[(N/2)!]^2}. \tag{32}$$

In Fig. 15 a, we report a comparison between the distribution (32) (dashed line) and a Poissonian with the same mean value (solid line). The difference between these distributions is below the resolution allowed by our experiment, as shown by the size of the error bars. In the case of a speckle pattern, because the vortices are generated by the linear superposition of uncorrelated events, namely the wavelets emitted by the scattering centers, the vortices are expected to have a Poisson distribution. Figure 15 b reports the collection of experimental points with their error bars, together with a Poissonian of the same mean value (solid line), which appears as a good fit.

Therefore, we conclude that the above statistical indicator is neither able to discriminate between the linear and the nonlinear case, nor in the second case, between a purely random distribution (Poissonian) and a specific creation and annihilation rate [GLM].

It has been theoretically predicted [CGL] and experimentally confirmed [GPR, Ar5] that vortices play a role in breaking the spatial correlation of the field. Indeed, it was shown [Ar5] that the mean nearest neighbor separation of vortices is close to the correlation length of the field. It is a relevant question to investigate whether vortices also have a role in breaking the time correlation of the system, once the observation is limited to a spatial correlation domain.

To answer this question, we measure the probability distribution of the time separation between successive onsets within a space correlation range both in nonlinear and linear cases. In the nonlinear experiment [Ar5], a correlation area of the field corresponds to a restricted frame of 35 x 35 pixels on the video camera. This area will contain, in general, a 0 or 1 vortex. We acquire a set of 12,000 frames separated in time by 0.2 sec. Then, we associate with these data a time series, assigning at each instant the number of vortices that are in the square at that time. A sketch of this time series $N(t)$ is shown in the inset of Fig. 16 a. By differentiating and considering only the points with positive derivative, we obtain the time series $\Delta N/\Delta t$ of the same inset (below), where each pulse corresponds to the entrance of a vortex into the spatial correlation box.

Because the sequence in the inset is a generic stripe of 200 frames, we can say that within a correlation area far away from the boundary annihilation and creation events are very rare as compared to motional effects (entrance and exit). Thus, the information provided by these time series is mainly related to vortex motion.

Using these sequences, we can build the probabilities $P(t)$ of having n successive events within a time t. Experimental results for $n = 0$ to $n = 3$ are shown in Fig. 16. The continuous curve is a Poissonian of the same mean value as the experimental data. The Poissonian would be the correct distribution of the data if the probability rate w(t) of occurrence (entrance in the square) of a vortex in the unit time was a constant, independent of t. Indeed, from $w(t) = w_0$, it follows

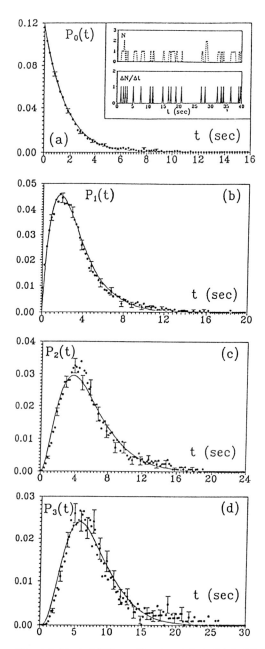

Fig. 16. Probabilities of having (a) 0, (b) 1, (c) 2, (d) 3 successive events within a time t, starting from a random initial time t_0; an event is an entrance of a defect into a correlation box and the data refer to the nonlinear experiment. For convenience, error bars have been reported every five points. Inset: Time sequences of (top) permanence of a defect in a correlation domain and (bottom) of the entrance of a defect in the same domain.

$$P_n(t) = \frac{(w_0 t)^n e^{-w_0 t}}{n!}.$$ (33)

However, the fit of the Poissonian to our data shows systematic deviations. In particular, the experimental values of P (t) for $n \geq 2$ are below the Poissonian for very short t and above the Possonian around the maximum of the curve, thus showing a sort of antibunching effect in the occurrence of vortices, compared to the randomly distributed time intervals implied by the Poissonian. This means that each vortex has an associated refractory time τ, so that the occurrence of one vortex reduces the probability of a next one in the successive instants.

Thus, we conjecture that if a vortex has entered the square at time t, then the rate of occurrence of a new vortex varies in time as

$$w(t) = w_0(1 - e^{-(t-t_0)/\tau}).$$ (34)

It is no longer possible to give an explicit form for $P_n(t)$ for $n > 0$, because it would be necessary to know the whole story of the evolution. Instead, we can make a prediction for the statistics $Q(t)$ of empty intervals between successive events. In effect, if one event has occurred at $t = 0$, according to (34), the probability that the next event will occur at $t_k = k\Delta t$ is

$$Q(t_k) = [1 - \int_0^{\Delta t} w(t)dt] \; ..... \; [1 - \int_{(n-2)\Delta t}^{(n-1)\Delta t} w(t)dt] \int_{(n-1)\Delta t}^{n\Delta t} w(t)dt$$

$$= [w_0\Delta t + w_0\tau[e^{-n\Delta t/\tau} - e^{-(n-1)\Delta t/\tau}]]$$

$$\cdot [\prod_{k=0}^{k=n-2} [1 - w_0\Delta t + w_0\tau(e^{-(k+1)\Delta t/\tau} - e^{-k\Delta t/\tau})]].$$ (35)

Comparing this distribution with the experimental data for the best fit that minimizes the mean square deviations, we get the values $\tau = 0.46$ s and $w_0 = 0.62 \; s^{-1}$. Notice that if we evaluate the correlation time of the sequence $N(t)$ (upper inset of Fig. 16a), we obtain a value close to τ, which then represents the average permanence time of a vortex within the observation area. In Fig. 17a, the experimental results are shown together with the best fit (solid line) of the distribution (35). The fit is seen to be good. The dashed curve in the same figure represents the Poissonian $P(t)$ (see (33)) with the same mean value.

We conclude that the arrival of a vortex in the square implies a refractory time equal to the average permanence time of the vortex in the square. Moreover, the arrival of each vortex induces a loss of correlation in the time series, because at each arrival time of a vortex, the probability $w(t)$ loses its memory.

We repeat the same measurement in the case of the speckle field, by taking 17,840 frames of 35×35 pixels at time intervals of 0.5 sec. In this case, because the diffuser is translated at a uniform speed, the statistics of the time between successive events corresponds simply to the statistics of distances between points on the diffuser that generate a vortex in the area of observation. In this case, the best fit of the distribution (35) with the data (solid line in Fig. 17b) yields

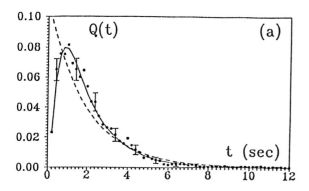

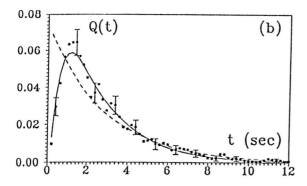

Fig. 17. Probability of having 0 successive events within a time t counted from an initial time t_0 corresponding to an event. The data refers to (a) the nonlinear case and (b) the linear case; the strong deviations from the dashed line (Poissonian fit) are due to the refractory time in the time series. Solid lines correspond to the best fit of (35) with the data.

$\tau = 0.61$ s and $w_0 = 0.45$ s^{-1}. This value of τ is trivially related to the uniform speed at which the diffuser is moving.

Going back to the nonlinear case, it has already been shown that the mean vortex separation $< D >$ in space and $< T >$ in time is related to the coherence range of the field (see Figs. 5 and 6 of [Ar5]). Here the mean separation time $< T >$ evaluated from Fig. 17a is 1.6 sec. Thus, the refractory time $\tau \simeq 0.4 < T >$ appears as a kind of "hard core" perturbation with respect to a model of point-like vortices. This suggests a possible spatial analogue.

2.4 Transition from Boundary- to Bulk-Controlled Regimes [Ar6]

Two types of pattern formation can arise in a system far from equilibrium, namely, the forced pattern, where the symmetry is imposed through the boundary, either by the geometry or by an external driving force; and the spontaneous pattern, where the symmetry is imposed by the bulk parameters of the active medium without any boundary influence.

Examples of the first type are convective instabilities in fluids, where the scale length is imposed by the cell height [Mnn] or the capillary pattern in fluid layers submitted to a periodic vertical force (Faraday instability) [EKR, FvK].

A Turing instability is the prototype of a bulk instability [Trn]. Indeed, let us consider a reaction-diffusion process with two competing species $i = 1, 2$ with diffusion lengths l_i, reacting through a nonlinear coupling. If we adjust the coupling so as to have an activator $(i = 1)$ and an inhibitor $(i = 2)$, then in the limit $l_1/l_2 \ll 1$ stationary patterns may arise with a scale length of the order of $\Lambda \sim (l_1 l_2)^{1/2}$, independent of the boundary [KrO]. Only recently [Cst] experimental evidence of chemical Turing patterns has been offered, because, beforehand, transport effects were stirring the chemical components, thus imposing a boundary dependence.

In this section, we report optical pattern formation for different "aspect ratios" (i.e., different Fresnel numbers) showing the transition from patterns dominated by the geometric parameters to patterns whose scale length is imposed by the bulk properties of the medium.

In the case of optical media, any device based on mode selection in a cavity much longer then the optical wavelength λ, yields transverse patterns depending on the Fresnel number. F includes the two relevant geometrical parameters, namely the transverse size a and the longitudinal size L of the cavity, and it accounts for competition between geometric acceptance and diffraction phenomena. As a result, patterns based on optical propagation are boundary dominated, even though they are sometimes called "Turing phenomena" [LgL]. Indeed, observed patterns in lasers can be explained exclusively in terms of symmetries imposed by the boundary [Grn].

Recent evidence of transverse optical patterns in a passive alkali cell showed hexagon formation, but the experiment was carried at fixed geometrical parameters [GLV] and no hints were offered on possible boundary-bulk trade off. Another experiment on an optical cavity with a dephasing slab of liquid crystals [McE] showed patterns scaling with the cavity length in a purely diffractive way.

In our experiment, a fundamental geometric parameter is the spot size of the central mode, which is constrained by the quasi-confocal configuration (intracavity lens of focal length $L/4$), positioned at a distance from the crystal close to $L/4$. Thus, the spot size of the central mode is given by [Kgl]

$$w_0 = \sqrt{\frac{\lambda L}{\pi}}. \tag{36}$$

Provided the mirror size a is larger than w_0, that is, that the Fresnel number $F = a^2/\lambda L$ is larger than 1, the cavity can house higher order modes, made of regular arrangements of bright spots (peaks of Gauss-Laguerre functions in cylindrical geometry) separated by

$$D \simeq \frac{w_0}{\sqrt{F}}. \tag{37}$$

Because the overall spot size of a transverse mode of order n scales as $\sqrt{n}w_0$, it is clear that $n = F$ is the largest order mode compatible with the boundary conditions (filling all the aperture area).

To check experimentally such a scaling law, we rely on the fact that, as discussed in Sect. 2.2, each intensity zero provides a phase singularity [Brr], which can be detected as a fringe dislocation by heterodyning with a tilted reference plane wave. In this way [Ar5], we provide a technique for measuring the average nearest neighbor separation $< D >$ between phase singularities, which represents the characteristic length of our pattern system. Notice that patterns built by superposition of Gauss-Laguerre functions have, in general, an average separation $< D >$ of zero intensity points approximately equal to the average separation D of bright peaks [Kgl].

A plot of $< D >$ versus F (Fig. 18a) shows that (37) is verified up to a critical value F_c, above which D is almost independent of F. In a similar way, the total number N of phase singularities scales as F^2 or F, respectively, below and above F_c. This transition is evident in Fig. 18b, and its root is the following. N is the ratio of the total wavefront area a^2, which scales as F, to the area $< D >^2$, containing a single defect, which scales as F^{-1} or F^0, respectively, below and above F_c.

We can understand the transition as follows. Consider the photorefractive crystal as a collection of uncorrelated optical domains, each one with a transverse size limited by a correlation length l_c intrinsic of the crystal excitations [Cm1]. Then the medium gain has un upper cutoff at a transverse wave number $1/l_c$, and spatial details are amplified only up to that frequency, that is, provided they are bigger than l_c. Thus, for a critical F_c, such that $D = \frac{w_0}{\sqrt{F_c}} = l_c$, we expect a transition from a boundary-dominated regime described by (37) to a bulk-dominated regime, whereby the separation of the phase singularities is independent of F. This is indeed the case as shown in Fig. 18, which yields a value $F_c \simeq 11$, corresponding to $l_c \sim 170$ μm, because $w_0 \sim 600$ μm for $L = 200$ cm.

The reduction of the boundary influence is also signaled by the reduction of the topological charge imbalance. Indeed, since a regular field should have a balance between topological charges of different signs [Brr], an imbalance means that two phase singularities of opposite signs have been created close to the boundary, and only one has remained within the boundary. Therefore, there is a boundary layer of area $a \cdot < D >$ containing $N_1 \sim a/ < D >$ singularities ($N_1 \ll N$ as soon as $< D > \ll a$) within which a topological charge imbalance can occur. The absolute value of the imbalance of positive to negative charges is, for statistical reasons, of the order of $\sqrt{N_1}$, and thus, the normalized imbalance is $U = \sqrt{N_1}/N$. Accounting for the scaling of $< D >$ with F, it follows that U scales as $F^{-1.5}$ and $F^{-0.75}$, respectively, below and above F_c. Figure 19 reports the experimental results, which are in agreement with this expectation.

To characterize l_c directly, we measure the single pass gain of BSO in a two-wave-mixing configuration without cavity, as shown in Fig. 20a. A lens provides an image of the near-field output on the video camera VD. A diaphragm on the focal plane of the lens filters out noise. By tilting the two waves, we measure

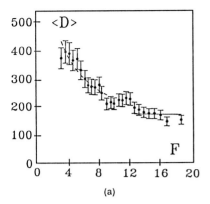

(a)

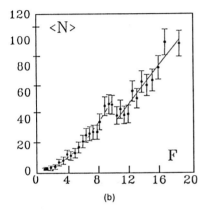

(b)

Fig. 18. (a) Mean nearest neighbor separation $< D >$ (scale in μm) between phase singularities. (b) Average total number $< N >$ of phase singularities versus the Fresnel number F of the cavity. Dashed lines are best fits with the boundary-dependent scaling laws $< D > \sim F^{-1/2}$ and $< N > \sim F^2$. Solid lines are best fits with the bulk-dominated scaling laws $< D > \sim F^0$ and $< N > \sim F^1$.

the photorefractive gain G_0 versus the grating spatial frequency K. As shown in Fig. 20b, we have the maximum gain for an optimal angle $\theta \sim 1°$, corresponding to a Bragg grating with periodicity $\tilde{K}^{-1} \sim \lambda/\theta \sim 30\ \mu$m. The solid line is a best fit according to the photorefractive theory of light scattering [GnH]. To measure l_c, we have to explore the broadening of the gain peak around $\tilde{K}$. When a diffuser is inserted on the signal path, it provides a spectral broadening q, controlled by translating the diffuser between two confocal lenses. In this way, we measure the amount of amplification of a speckle field of varying size. The data (Fig. 20c) yield a cutoff at $q_c \sim 5$ mm^{-1}, whose reciprocal is in agreement with the indirect evaluation of l_c.

The work of D'Alessandro and colleagues [DAF] also predicts a transition from a purely dispersive pattern length $(\lambda L)^{1/2}$ to a partly dissipative length $(\lambda L l_c)^{1/3}$. This partly dissipative length, however, still contains the geometric

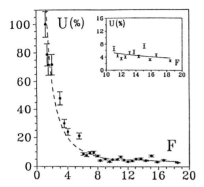

Fig. 19. Average charge unbalance $U \equiv <\mid n_+ - n_- \mid /(n_+ + n_-) >$ versus F. Here $n_+(n_-)$ is the total number of positive (negative) phase singularities as defined in [CGL]. Dashed line $F^{-1.5}$ fit up to $F = 11$, solid line $F^{-0.75}$ fit from $F = 11$ on. The inset provides an expanded view of the $F > 11$ region.

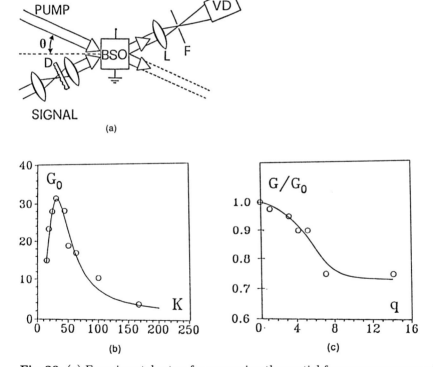

Fig. 20. (a) Experimental setup for measuring the spatial frequency response of BSO. D: diffuser, L: lens, and F: diaphragm. (b) Gain G_0 (ratio of the output signal intensities with and without pump) versus K (measured in mm^{-1}). (c) Gain G versus spectral broadening $q(\text{mm}^{-1})$, measured with θ fixed at the optimum value. The solid line is a guide for the eye.

parameter L, and for sufficiently small L, it would provide a pattern length even smaller than l_c. In fact, an intrinsic cutoff length does not emerge from that treatment, because saturation effects of the nonlinearity are completely neglected.

In the set of data reported in Sect. 2.2, the range of F numbers explored was not wide enough to give evidence of two separated regimes, thus we fit the available data for $< D >$ and N versus F with global exponents slightly different from -0.5 and 2 (-0.62 and 1.79, respectively). The present extension to larger F numbers allows clear separation of the two regimes.

Finally, above F_c, the persistence time $< T >$ of a phase singularity within a domain of size $< D >$ also has to be independent of F. This was evident in Fig. 6a of [Ar5], although no adequate explanation could be offered at that time.

A further independent check of bulk-dominated patterns is given by the power spectrum $S(q)$ of the transverse optical field for $F > F_c$. This measurement is done by integrating the signal intensity over concentric shells of radius q in the Fourier space provided by the far-field propagation of the cavity field. The results are reported in Fig. 21. The best fit of the data yields an exponential high-frequency cutoff e^{-q/q_0} with $q_0 = 5.3$ mm^{-1}. This corresponds to a correlation length $1/q_0 \sim 190$ μm for the field, which is in good agreement with the values of l_c reported above. A broadband spectrum with an exponential cutoff is a clear signature of spatio-temporal chaos [PPP, HhS], where a chaotic dynamics of spatial patterns are involved with a dominant length scale.

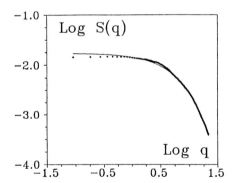

Fig. 21. Field power spectrum measured integrating the signal intensity over concentric shells of radius $q(\text{ mm}^{-1})$ in the Fourier space. Dots are experimental points, $F \equiv 70$. Solid line – best fit with $\exp[-(q/q_0)^\beta]$, yielding $\beta = 0.96$, $q_0 = 5.3$ mm^{-1}.

The transition to a boundary-independent regime reported here can be explained as follows. Linearizing the dynamics at the onset of the instability, the gain per unit length $g(q)$ provided by the crystal for each q component should be equal to the attenuation $\alpha(q)$, to have marginal stability.

The attenuation $\alpha(q)$ is derived by the following considerations. A spectrophotometric measurement gives a transmission $T_1 = 37\%$ through the crystal in the absence of a pump, corresponding to a linear attenuation α_1. Furthermore, the optical intracavity elements (beam splitter, lenses) add an attenuation $1 - T_2 = 20\%$. Reducing this last one to an equivalent crystal attenuation α_2, we have $\exp[-(\alpha_1 + \alpha_2)l] = T_1 T_2 = 0.296$, yielding $\alpha = \alpha_1 + \alpha_2 = 1.2$ cm^{-1}, because the crystal length is $l = 1$ cm. Beside this uniform attenuation there is a q dependent attenuation $\alpha_d(q)$ due to the diffractive spread of the beam, so that the overall attenuation is $\alpha(q) = \alpha + \alpha_d(q)$. A beam of radius r at the crystal exit, reenters after a cavity round trip of length L, with a radius increased by diffraction. The q component of the field (diffraction angle $\theta(q) = q/k$, where q and $k = 1/\lambda$ are the lengths of the transverse and longitudinal wavevectors), will be spread over a radius $r + Lq/k$, thus the relative loss due to the limited crystal aperture is

$$1 - e^{-\alpha_d(q)l} = \frac{(r + Lq/k)^2 - r^2}{(r + Lq/k)^2}. \tag{38}$$

From this relation, $\alpha_d(q)$ can be derived for different Fresnel numbers $F = r^2/\lambda L$. In Fig. 22, we report the marginal equilibrium curves $\alpha(q) = \alpha + \alpha_d(q)$ versus q. In that plot, we also report the actual gain per unit length $g(q)$ obtained from the data in Fig. 20c. Indeed, Fig. 20b reports the ratio of the transmission in axis with and without a pump, which amounts to $G_0 = \exp(g_0 l)$, thus yielding $g_0 = 3.5$ cm^{-1}. An observer placed at a propagation angle q/k detects a reduced gain $G(q) = \exp(g(q)l)$, as plotted in Fig. 20c, from which we deduce $g(q)$.

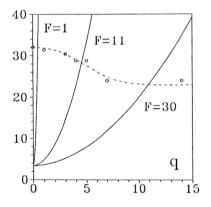

Fig. 22. Marginal stability curves versus q for different Fresnel numbers (solid lines) and gain curves (dashed line). For convenience, we have reported the global gain $G(q)$ and the inverse transmission $T^{-1}(q)$, rather than $g(q)$ and $\alpha(q)$.

As can be seen in Fig. 22, for low Fresnel numbers, the gain curve is practically flat, therefore, the mode selection depends exclusively on $\alpha(q)$, that is, on the boundary symmetries [Ar7], thus, it is irrelevant to worry about a detailed material model. On the contrary, for $F > 10$, the limited half-width

$q_c = 1/l_c \sim 5 \ \mathrm{mm}^{-1}$ of the gain line constrains the minimum spot size, independently of the maximal q allowed by the marginal stability curve. In fact, while the linear stability analysis enables all modes within the marginal curve to oscillate, the nonlinear coupling selects those close to the gain peak, that is, those confined within the half-width of the gain curve.

3 Theory of Pattern Formation and Pattern Competition

3.1 Equations of Photorefractive Oscillator [Ar8]

Two different models are currently used to describe light-induced charge transport in photorefractive materials – the band transport model by Kukhtarev and colleagues [Kkh] and the hopping model by Feinberg and colleagues [Fnb]. In fact, both models provide essentially equivalent descriptions of the process [JHF], although they use a different language and a different set of physical parameters. Nevertheless, some debate on the relative virtues of each model is still continuing [Krn]. In the Feinberg model, carriers are assumed to experience jumps among trap sites, with a rate that is a function of the distance. The band transport model involves coherent motion of the photo-excited carriers between scattering events by lattice imperfections and phonons. This model is more generally used and is better accommodated in the standard analysis of carrier migration in insulating and semiconductor materials.

Kukhtarev's approach has been used in a recent model of a photorefractive ring oscillator [DAl]. This model, however, leads to stationary solutions, and it offers no evidence of PA and CA. Furthermore, available experimental data to be fit into Kukhtarev's model are widely scattered and sensitive to the measuring procedures used [ACA].

On the other hand, the interaction with the optical field implies the collective properties of the photorefractive medium, which are characterized by a susceptibility field $\chi(r, t)$. We thus propose to build a phenomenological model of the χ dynamics, in terms of collective parameters that can be measured by suitable laboratory experiments.

We aim to build a specific model that contains all the relevant physical aspects of the experiment described above, without resorting to unnecessary microscopic assumptions. The requirement is that all the relevant parameters be characterized by the experimental setup, plus the addition of simple optical measurement of bulk properties done on the crystal without cavity [Ar6].

The two-wave mixing gives rise to a grating (dotted lines), which breaks the cylindrical symmetry of the cavity. The grating spacing is $\Lambda = \frac{1}{k_0\theta}$, with k_0 the wave number of the pump, and $\theta \ll 1$ the angle between the two waves, which in the experiment [Ar4] is optimized at $\theta \simeq 1/60$ rad.

Introducing an x, y, z coordinate system with z along the cavity axis, with $x - z$ being the plane that contains the propagation directions of the two fields E_p and E, and with the origin at the intersection of the input face of the crystal with the cavity axis, then the equation for the total field $\tilde{E} = E_p + E$ is

$$\sqcap^2 \tilde{E} = -\mu \partial_t^2 P \ \text{rect}(z/l), \tag{39}$$

where $\text{rect}(z/l) \equiv 1$ for $0 < z < l$ and $\text{rect}(z/l) \equiv 0$ elsewhere, with l being the crystal thickness. The derivative signs are abbreviated as e.g., ∂_t^2 for ∂^2/∂_t^2, etc.

Because the cavity length L is much longer than l, the source term on the right side is strongly localized. The polarization is given by

$$P = \varepsilon_0 \chi \tilde{E}, \tag{40}$$

where the susceptibility field $\chi(\mathbf{r}_\perp, t)$ $(\mathbf{r}_\perp = (x, y)$ (stays for the transverse coordinate) has no z dependence $(l \ll L)$ and obeys the equation

$$(\partial_t + \frac{1}{\tau} - D\nabla_\perp^2)\chi = f(\tilde{E}), \tag{41}$$

where $\nabla_\perp^2 \equiv \partial_x^2 + \partial_y^2$.

Equation (41) arises from the following considerations. The two-wave mixing gives rise to a space charge field $\chi(r, t)$, consisting of a dipole distribution created by the offset of the free electrons with respect to the fixed donors in the BSO material. The collective wave χ decays locally with a lifetime τ and diffuses at a rate D. While D corresponds to the currently available data for diffusion of free electrons, τ is a phenomenological parameter that has nothing to do with the so-called recombination time $\tau_R \sim 10^{-5}$ sec of free electrons [ACA]. In fact, here τ is the damping time of the collective dielectric excitation $\chi(r, t)$, and direct measurement on the crystal used provides a value $\tau \simeq 1$ sec [Ar6].

To solve (39) – (41), we first isolate a fast varying part in z and t from the slow space-time dependence, which is relevant for the dynamics, by writing

$$\tilde{E} = E_p e^{-i(\mathbf{k}_0 \cdot \mathbf{r} - \omega_0 t)} + E(\mathbf{r}_\perp; z, t)e^{-i(kz - \omega t)}. \tag{42}$$

This way, we have extracted from the cavity field a plane wave along the cavity axis (k along z), thus E includes the full dependence on $\mathbf{r}_\perp$, plus the residual slow dependence on z and t.

The applied dc field E_{dc} in the x direction induces a space-charge drift, with a speed $v_d = \mu E_{dc}$ (μ = electron mobility), thus the scattering grating provides a frequency offset along x given by

$$\Omega = v_d k_0 \theta. \tag{43}$$

Consequently, the cavity field has a frequency ω detuned with respect to the pump frequency ω_0, by

$$\omega = \omega_0 - \Omega. \tag{44}$$

By use of (42) and neglecting the second "slow" derivatives (so-called eikonal approximation), the $\sqcap^2$ kernel reduces to $\partial_t + c\partial_z - \frac{ic}{2k}\nabla_\perp^2$. Furthermore, we approximate $\partial_t^2 P \approx -\omega^2 P$. By Fourier expansion of the transverse dependence

$\mathbf{r}_\perp \rightarrow \mathbf{k}$, the product $\chi \tilde{E}$ of (40) transforms into a convolution. In this one, only the two components

$$\chi_0 \equiv \chi(\mathbf{k} = 0)$$

and

$$\chi_1 \equiv \chi(-\mathbf{K} = \mathbf{k} - \mathbf{k}_0),$$

multiplied, respectively, by the cavity and pump fields, yield phase-matched contributions to (40). Equation (39) becomes precisely

$$(\partial_t + c\partial_z - \frac{ic}{2k}\nabla_\perp^2)E = -\frac{i\omega}{2}(\chi_0 E + \chi_1 E_p) \quad \mathtt{rect}(z/l). \qquad (45)$$

The material in (41) provides two separate equations for the two Fourier components χ_0 and χ_1.

The transverse Fourier expansion has provided the $\mathbf{k} = 0$ and $\mathbf{k} = \pm\mathbf{K}$ components of the χ grating, which scatters the two fields E_p and E into one another. Only the components χ_0 and χ_1 (see Fig. 23) are coupled with the cavity direction.

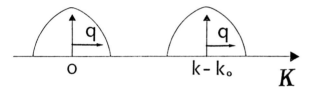

Fig. 23. Bragg peaks in two-wave mixing and broadening $\Delta K = q$ due to diffusive processes in the crystal.

By the same argument of separation of fast and slow variables in the longitudinal dependence used in going from (39) to (45), the Fourier spectral components χ_0 and χ_1 account for the "fast" space variations on a scale

$$\frac{2\pi}{K} \equiv \Lambda \simeq \frac{\lambda}{\theta} \simeq 30\mu m,$$

($\theta \sim 1/60$, $\lambda \sim 0.5\ \mu m$).

The residual transverse "slow" space dependence, on a length scale comparable with the wavefront size, is included in the differential terms. Furthermore, the component χ_1 is oscillating at the frequency $\Omega = \omega - \omega_0$, which can be put into evidence by the transformation

$$\chi_1 \Longrightarrow \chi_1 e^{i\Omega t},$$

which means to replace ∂_t by $\partial_t + i\Omega$.

By the above scale consideration, we can rewrite the equation for the χ field as

$$(\partial_t + \frac{1}{\tau} - D\nabla_\perp^2)\chi_0 = f_0(\tilde{E}), \qquad (46)$$

$$(\partial_t + i\Omega + \frac{1}{\tau} - D\nabla_\perp^2)\chi_1 = f_1(\tilde{E}). \qquad (47)$$

Notice that the parameters τ, Ω, and D, appearing on the left side can be measured by a macroscopic optical experiment on the crystal [Ar6, Rmz] without a microscopic knowledge of the solid-state phenomenon peculiar to photorefractive media. Indeed, the wave number $\mathbf{K}$, at which the Bragg scattering line has to be positioned, was measured by optimizing the photorefractive gain in a single pass amplification [Ar6].

The broadening Δk around that peak (Fig. 20b) measured by imaging speckle fields of different sizes, yields a correlation length $l_c = 2\pi/\Delta k$ intrinsic to the material. However, because the main source of decorrelation is the free charge diffusion away from the interference maxima, where it is created, we can consider l_c as a diffusion length over the lifetime τ of the collective space-charge wave. Thus we evaluate a diffusion coefficient

$$D = \frac{l_c^2}{\tau}. \qquad (48)$$

Indeed, the values $l_c \sim 200$ and μm, $\tau \sim 1$ sec measured in [ACA, Rmz] provide for the BSO crystal used in the experiments, with a value $D \sim 4 * 10^{-4}$ cm^2/sec, within the range of values reported in the literature [ACA].

To characterize the source terms $f_0(\tilde{E})$, $f_1(\tilde{E})$ of the χ equation, it is necessary to build a suitable model of the $\chi - E$ interaction.

Without entering into microscopic details on the photorefractive effect, the build-up of a χ amplitude can be modeled as the annihilation of a pump photon with the simultaneous creation of a cavity photon and a quantum of the space-charge wave.

If we consider the scattering process as an interaction among three Bose fields, described by the operators a_p, a_p^+ for the pump field E_p of wavevector $\mathbf{k}_0$; a, a^+ for the cavity field E of wavevector $\mathbf{k}$, and b, b^+, for the space-charge field, giving the polarization component χ_K ($\mathbf{K} = 0$ and $\mathbf{K} = \mathbf{k}_0 - \mathbf{k}$), the dynamics are accounted for by an effective Hamiltonian interaction

$$\frac{H_{eff}}{\hbar} = ig(a_p a^+ b^+ - h.c.), \qquad (49)$$

together with the Hamiltonian for the free fields

$$H_0/\hbar = \omega_p a_p^+ a_p + \omega a^+ a + \Omega b^+ b, \qquad (50)$$

where $\Omega = \omega_p - \omega$ is the frequency of the space-charge mode excited by the scattering process from the pump to the cavity field. In the presence of an applied dc field E_{dc} across the crystal, along a direction parallel to $\mathbf{K}$, the grating of wavevector $\mathbf{K}$ undergoes a translation at speed $v = \mu E_{dc}$ (where μ is the electron mobility), thus the Doppler shift provides a detuning $\Omega = kv$.

If we now introduce the notation

$$\sigma^- = a_p b^+, \quad \sigma^+ = a_p^+ b$$

$$\sigma_3 = a_p^+ a_p - b^+ b, \tag{51}$$

it is easily seen that the σ operators obey Pauli commutation rules. Indeed, this is the well known Schwinger representation for pairs of Bose operators [Mss].

Because the Hamiltonian (49)+(50) implies the conservation of the total number of pump and signal photons

$$n_p + n_s = a_p^+ a_p + a^+ a, \tag{52}$$

the total Hamiltonian can be rewritten (except for constant terms that do not contribute to the equations of motion) as

$$\frac{H}{\hbar} = \frac{\omega}{2}\sigma_3 + \omega a^+ a - ig(\sigma^+ a - \sigma^- a^+). \tag{53}$$

Equation (53) is the standard Hamiltonian for a two-level atom coupled to a single-mode radiation field.

The corresponding Heisenberg equations of motion are

$$\dot{a} = -i\omega a + g\sigma^-,$$

$$\dot{\sigma}^- = -i\omega\sigma^- + g\sigma_3 a, \tag{54}$$

$$\dot{\sigma}_3 = -2g(\sigma^+ a + \sigma^- a^+).$$

We now consider the coupling of this microscopic system with a heath bath, giving rise to dissipation terms and destroying the quantum correlations. Such a heath bath accounts for all dissipative processes occurring in a real crystal.

The previous equations can be rewritten as C-number equations for the corresponding averages over the thermal fluctuations. Keeping the same notation $a, \sigma^\pm, \sigma_3$ for $< a >, < \sigma^\pm >, < \sigma_3 >$ and referring to a frame rotating at the common angular velocity ω, we have

$$\dot{a} = g\sigma^- - ka,$$

$$2\dot{\sigma}^- = g\sigma_3 a - \gamma_\perp \sigma^-, \tag{55}$$

$$\dot{\sigma}_3 = -2g(\sigma^+ a + \sigma^- a^+) - \gamma_\parallel(\sigma_3 - \sigma_0).$$

Here $k, \gamma_\perp, \gamma_\parallel$ are the decay rates for the field a, with σ^- and σ^+ considered as C-number quantities.

We recall [Ar4, Ar5] that the cavity field decays over 10^{-7} sec, hence $k = 10^7 \ s^{-1}$, the pump field crosses the crystal without confinement mirrors, thus $\gamma_\perp \simeq \gamma_\parallel \sim 10^{10} \ s^{-1}$ (reciprocal of the transit time across the crystal thickness).

Such a numerical assignment is obvious for $\gamma_\parallel$, because σ_3 is proportional to the photon number $a_p^+ a_p$. As for $\sigma^\pm$, the phases of their complex factors decrease over the same time. Thus $\gamma_\perp$ is of the same order of $\gamma_\parallel$.

Due to the difference in the dissipative rates, we can adiabatically eliminate the second and third equations and replace the stationary solutions into the first of (55). Hence we arrive at the following equation:

$$\dot{a} = -ka + \frac{g^2}{\gamma_\perp} \frac{\sigma_0 a}{1 + \frac{|a|^2}{|a_s|^2}}. \tag{56}$$

Here, the pump term σ_0 is practically coincident with the pump intensity, because $a_p^+ a_p \gg b^+ b$, always.

The saturation term is given by

$$|a_s|^2 = \frac{\gamma_\perp \gamma_\parallel}{g^2}. \tag{57}$$

In conclusion, (56) provides an evolution equation for the signal field within the medium as

$$\dot{E} \propto \frac{|E_p|^2 E}{1 + |E|^2 / |E_s|^2}. \tag{58}$$

If $|E| \ll |E_s|$, (51) becomes

$$\dot{E} \propto |E_p|^2 E(1 - \frac{|E|^2}{|E_s|^2}),$$

and normalizing all the fields to the value of E_s ($E_p \rightarrow E_p/E_s$ and $E \rightarrow E/E_s$)

$$\dot{E} \propto |E_s|^3 |E_p|^2 E(1 - |E|^2). \tag{59}$$

The above consideration has been developed for three pure modes, that is, for three plane waves interacting.

Recalling that the source term for the evolution of E should be proportional to $P = \varepsilon_0 \chi \tilde{E}$, and that the asymptotic solution of (46, 47) (for a very slow field) would be of the type

$$\chi_0 = K_0 f_0(\tilde{E}),$$
$$\chi_1 = K_1 f_1(\tilde{E}), \tag{60}$$

where K_0 and K_1 are suitable linear kernels, we can identify f_0 and f_1 with the following expressions:

$$f_0(\tilde{E}) = r |E_p|^2 (1 - |E|^2),$$
$$f_1(\tilde{E}) = r E_p^* E(1 - |E|^2) \tag{61}$$

(r is an opportune coefficient) to be replaced in (46, 47).

3.2 Truncation to a Small Number of Modes: Numerical Evidence of PA, CA, and STC

The coupled (45) and (46, 47) with the source terms (61) describe the dynamics of the photorefractive oscillator. First, we notice that (45) and (46-47) imply two very different time scales. Integration of (39) on z implies the right-hand source term on a thin slice $l \ll L$, plus a free propagation along the rest of the cavity of length $L - l \sim L$, including reflection losses on the mirrors. Because $L \simeq 2m$, the transit time on the ring, takes a few nanoseconds. Furthermore, because the intracavity losses for forward propagation are very high (BSO has a transmission of 37%), the overall cavity damping time t_c is shorter than 100 nsec. On the contrary, the material excitation χ decays over a time scale $\tau \sim 1$ sec. Let us introduce a slow-time scale

$$T = \frac{\tau}{t_c} t = 10^7 t. \tag{62}$$

Furthermore, we introduce nondimensional transverse coordinates normalized to the beam waist $w_0 = \sqrt{\frac{\lambda L}{2\pi}}$

$$x \rightarrow \frac{x}{w_0} \qquad y \rightarrow \frac{y}{w_0}, \tag{63}$$

and take the transverse derivatives with respect to them. The equations then become

$$(10^{-7}\partial_T + c\partial_z - \frac{ic}{2kw^2}\nabla_\perp^2)E = -i\frac{\omega}{2}(\chi_0 E + \chi_1 E_p)\texttt{rect}(z/l), \tag{64}$$

$$(\partial_T + \frac{1}{\tau} - \frac{D}{w^2}\nabla_\perp^2)\chi_0 = r \mid E_p \mid^2 (1 - \mid E \mid^2), \tag{65}$$

$$(\partial_T + i\Omega + \frac{1}{\tau} - \frac{D}{w^2}\nabla_\perp^2)\chi_1 = rE_p^* E(1 - \mid E \mid^2). \tag{66}$$

We now approximate the three partial differential equations by a finite set of ordinary differential equations.

The solutions of the free propagation problem (64) (without source) with a cylindrical symmetry (round boundary in x,y) are a set of Gauss-Laguerre eigenfunctions [Kgl]

$$E_{p0}(\rho, \phi) = \frac{2}{\sqrt{2\pi}}L_p^0(2\rho^2)e^{-\rho^2},$$

$$E_{pl1}(\rho, \phi) = \frac{2}{\sqrt{\pi}}(2\rho^2)^{l/2}(\frac{p!}{(p+l)!})^{1/2}L_p^l(2\rho^2)e^{-\rho^2}cos(l\phi), \tag{67}$$

$$E_{pl2}(\rho, \phi) = \frac{2}{\sqrt{\pi}}(2\rho^2)^{l/2}(\frac{p!}{(p+l)!})^{1/2}L_p^l(2\rho^2)e^{-\rho^2}sin(l\phi),$$

where $\rho = \sqrt{(x^2 + y^2)}/w_0$, $p = 0, 1, ...$ is the radial index, $l = 0, 1, ...$ is the angular index, L_p^l are the Laguerre polynomials of indicated arguments, and there is the supplementary condition for the radial and angular indices

$(2p+l)_{\texttt{max}} = F$. The Gauss-Laguerre functions obey the following orthonormalization rules:

$$\int_0^{2\pi} d\phi \int_0^\infty \rho d\rho E_{pli}(\rho, \phi) E_{p'l'i'}(\rho, \phi) = \delta_{pp'} \delta_{ll'} \delta_{ii'} \qquad (i, i' = 1, 2). \qquad (68)$$

From the above condition on the indices of the Gauss-Laguerre modes, adjustment of the Fresnel numbers acts as a low-pass filter, which cuts off high-order modes.

First, we consider an expansion limited to the three lowest order modes. Notice that rather than referring to a particular set of eigenfunctions supported by the boundary symmetry of the free propagation problem, we could have performed a generic Fourier expansion of the residual transverse (x, y) dependence into a $\mathbf{q}$ variable.

This $\mathbf{q}$ transform accounts for the slow space variations. These yield a broadening around $\mathbf{k} = 0$ and $\mathbf{k} = \pm\mathbf{K}$.

As shown in Fig. 2, lens L' has a focal length $L/4 \simeq 10^2$ cm. Thus, the folded ring cavity with four plane mirrors at the corners is, in fact, a quasi-confocal cavity. The crystal is positioned at the focus of L', where the spot size is $w_0 = \sqrt{\frac{\lambda L}{2\pi}} \sim 4 * 10^{-2}$ cm. Because the size of the transverse crystal is $a_c \simeq 0.5$ cm, the maximum order of transverse modes accepted by the crystal boundary (acting as a diaphragm) is

$$n_{\texttt{max}} \equiv F = \frac{a_c^2}{w_0^2} \simeq 140, \qquad (69)$$

because the nth order mode has a radium $\rho_n = \sqrt{n} w_0$. The total number of accepted modes scales as F^2.

Adding a variable pupil as in [Ar4], $n_{\texttt{max}}$ is then limited to about $F = 100$, which is within the above limitation due to the crystal size.

Because a mode of order n has n^2 peaks, the average size of each bright spot is $w_n = w_0/\sqrt{n}$. However, when this size is smaller than the correlation length $l_c = \sqrt{D\tau}$ intrinsic of the material, diffusion destroys the high-frequency grating necessary to build up modes of higher order. This transition from a boundary-dominated to a bulk-dominated regime was shown in Sect. 2.4. The transition occurs at a critical $F_c \simeq 11$ [Ar6], beyond which the size of the spots cannot decrease below l_c.

Adjusting the pupil aperture to select only three modes, we can expand the three fields E, χ_0, and χ_1 in a set $|i> (i = 1, 2, 3)$, where $|i>$ is the first three Gauss-Laguerre modes.

The transverse expansions are then

$$\begin{aligned}
E &= a \,|\, 1> +b \,|\, 2> +c \,|\, 3>, \\
\chi_o &= t_a \,|\, 1> +t_b \,|\, 2> +t_c \,|\, 3>, \\
\chi_1 &= (s_a \,|\, 1> +s_b \,|\, 2> +s_c \,|\, 3>)e^{-ik_0\theta w\rho \cos\varphi}.
\end{aligned} \qquad (70)$$

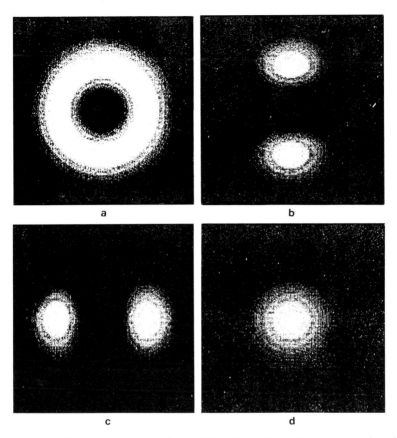

Fig. 24. Pattern reconstructions of the three Gauss-Laguerre modes chosen for the fields expansion (b, c, d) and the doughnut-shaped (a) superposition of the two Gauss-Laguerre modes of order 1.

While the transforms of E and χ_0 account for the residual slow variations around $\mathbf{k} = 0$, the transform of χ_1 has to include a "carrier" of wave number $-K = -k_0\theta$, which provides a common phase factor

$$e^{-i\mathbf{K}\cdot\mathbf{r}_\perp} = e^{-ik_0\theta w\rho\cos\varphi}. \tag{71}$$

The partial differential equations (64 - 66) become a set of nine ordinary differential equations for the nine complex amplitudes $a, b, ..., s_c$.

The fact that the experiments at low Fresnel numbers show evidence of both $\mid 2 >$ and $\mid 3 >$ modes means that the cylindrical symmetry is partially restored by the diffusion term as we will show later.

Because of the different scales of dissipation times, we can adiabatically eliminate the fast variables a, b, c and replace their equilibrium values in the remaining six equations for $t_a, ...s_c$. The three field equations, integrated over z, provide a

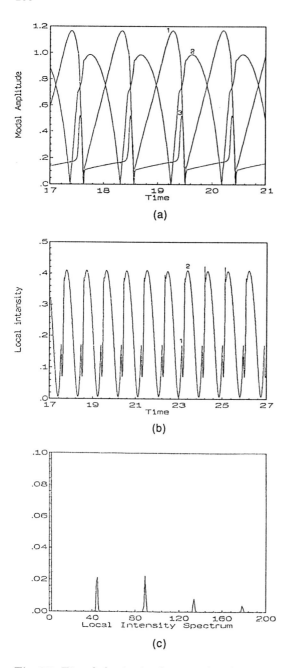

Fig. 25. Time behavior in the periodic alternation case (simulation). (a) Modal amplitude of three Gauss-Laguerre modes versus time (periodic alternation). (b) Local intensity measured in one point of the observation window versus time. (c) Spectrum of the local intensity.

mapping from $z = 0$ to $z = L$; on the interval $z = 0$ to $z = l$, we have the source term, while from $z = l$ to $z = L$, we have free propagation.

Figures 24, 25, and 26 show the numerical solutions of the modal equation, periodic alternation and chaotic alternation among the three modes considered above that have been observed for different values of the parameters.

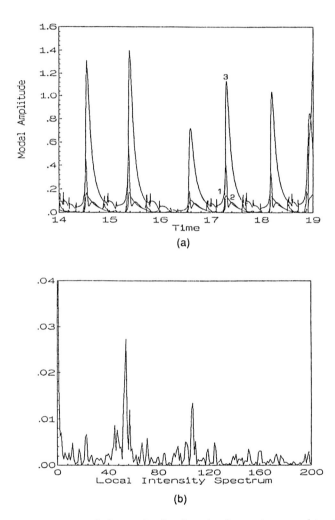

Fig. 26. Time behavior in the chaotic alternation case (simulation). (a) Modal amplitude of the three modes versus time. (b) Spectrum of the local intensity measured in a point of the observation window.

3.3 Symmetry Breaking at the Onset of Pattern Competition [Ar7]

Let us consider a simple dynamics involving three transverse modes, a central one with amplitude z_0 and two higher order modes counter-rotating along the azimuthal coordinate θ with respective amplitudes z_1 and z_2 and angular momenta ± 1. We can expand the cavity field as

$$E = f_1(r,l)(z_1 e^{i\theta} + z_2 e^{-i\theta})e^{i\omega_1 t} + f_0(r,l)z_0 e^{i\omega_0 t}, \tag{72}$$

where f_0 and f_1 are the space distributions of the modes. The optical frequencies ω_0 and ω_1 are, in general, different, and slow-time dependence due to the dynamics is included in the amplitudes $z_i(t)$ $(i = 0,1,2)$.

The zero-intensity situation corresponds to $z_0 = z_1 = z_2 = 0$, the central mode to $z_1 = z_2 = 0$, and an azimuthal standing wave to $z_0 = 0, z_1 = z_2$.

The time sequence of these three situations is one of the simplest cases observed experimentally [Ar4], so we aim to build model equations having the above sets of z values as fixed points. However, because any quasi-stationary point persists for a finite time, each of the fixed points must have at least an unstable direction. With these general rules in mind, we now discuss the symmetry requirements.

The observed symmetries impose the following constraints on the mode amplitudes [GSS]:

$$\Theta : (z_1, z_2, z_0) \rightarrow (e^{i\theta}z_1, e^{-i\theta}z_2, z_0),$$

$$K : (z_1, z_2, z_0) \rightarrow (z_2, z_1, z_0),$$

where Θ denotes the rotation operation and K the reflection around the privileged plane. As these modes are born by Hopf bifurcations, there is an additional time symmetry

$$B : (z_1, z_2, z_0) \rightarrow (e^{i\beta_1}z_1, e^{i\beta_1}z_2, e^{i\beta_0}z_0).$$

The normal form for the nonlinear interaction among the three modes, assuming that the symmetry of the system is Z_2 (reflection) degenerated toward an $O(2)$ one (reflection and rotation), is [DnK, Cm2] (dots denote time derivatives)

$$\dot{z}_0 = \lambda_0 z_0 + (a(|z_1|^2 + |z_2|^2) + b|z_0|^2)z_0,$$
$$\dot{z}_1 = \lambda_1 z_1 + (c|z_1|^2 + d|z_2|^2 + e|z_0|^2)z_1 + \varepsilon z_2, \tag{73}$$
$$\dot{z}_2 = \lambda_1 z_1 + (d|z_1|^2 + c|z_2|^2 + e|z_0|^2)z_2 + \varepsilon z_1,$$

where $\lambda_0, \lambda_1, a, b, c, d, e$ are complex coefficients and $\varepsilon = \rho_\varepsilon e^{i\varphi_\varepsilon}$ is a symmetry-breaking parameter. Letting $z_i = \rho_i e^{i\varphi_i}$, and changing the variables as $\rho_1 = A\cos(\alpha/2), \rho_2 = A\sin(\alpha/2)$ and $\delta = \varphi_2 - \varphi_1$, (73) will read

$$\dot{A} = (\lambda_1^r + (c^r - (1/2)(c^r - d^r)\sin^2(\alpha))A^2)A + (\rho_\varepsilon \sin(\alpha)\cos(\delta)\cos(\varphi_\varepsilon)$$
$$+ e^r \rho_0^2)A,$$
$$\dot{\alpha} = -\frac{1}{2}(c^r - d^r)\sin(2\alpha)A^2 + 2\rho_\varepsilon(\cos(\varphi_\varepsilon)\cos(\delta)\cos(\alpha)$$

$$+ \sin(\varphi_\varepsilon)\sin(\delta)),$$

$$\dot{\delta} = -(c^i - d^i)A^2 \cos(\alpha) + \rho_\varepsilon(\cotan(\frac{\alpha}{2})\sin(\varphi_\varepsilon - \delta)$$

$$-tg(\frac{\alpha}{2})\sin(\varphi_\varepsilon + \delta)),$$

$$\dot{\rho}_0 = (\lambda_0^r + a^r A^2 + b^r \rho_0^2)\rho_0, \tag{74}$$

and

$$\dot{\varphi}_0 = \lambda_0^i + a^i A^2 + b^i \rho_0^2,$$

$$\dot{\varphi}_1 = \lambda_1^i + A^2(c^i \cos^2(\frac{\alpha}{2}) + d^i \sin^2(\frac{\alpha}{2})) + e^i \rho_0^2. \tag{75}$$

We will now show that the solutions of (74, 75) reproduce the experimental behavior for certain parameter values that we derive explicitly. Notice that (74) constitutes a closed four-dimensional system.

Both the laboratory experiment [Ar4] and the numerical solution of the physical model [Kkh] show that an initial condition close to a central mode is followed in time by a zero intensity state (see Fig. 27). Therefore, in the six-dimensional phase-space of the solutions of (74, 75), the zero is stable in the ρ_0 direction. In our model, a first condition for the correspondence with the experiments is thus, $\lambda_0^r < 0$ and $b^r > 0$.

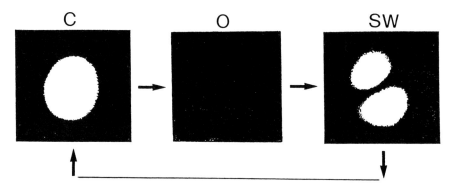

Fig. 27. Experimental sequence of three alternating patterns. C is the central mode (only $z_0 \neq 0$), O is the zero mode ($z_0 = z_1 = z_2 = 0$), SW (standing wave) is a balanced combination of clockwise (z_1) and counterclockwise (z_2) azimuthal traveling waves. The two spots of SW are aligned along the intersection with the privileged plane, corresponding to $\delta = 0$. Each of the three patterns lasts around 100 sec and is replaced by the next pattern within 2 sec, along the sequence shown by the arrows.

If $\rho_0 = 0$, there are fixed points at $\alpha = \pi/2, \delta = 0, \pi$, with

$$A^2 = -2(\lambda_1^r \pm \rho_\varepsilon \cos(\varphi_\varepsilon))/(c^r + d^r). \tag{76}$$

These solutions correspond to standing waves, and, therefore, will be part of the "skeleton" of pure modes in our dynamical model.

The stability of these fixed points in the (α, δ) directions can be derived from the eigenvalues of the Jacobian of the first three (74)

$$
\begin{bmatrix}
-2\lambda_1^r \mp 2\rho_\varepsilon \cos(\varphi_\varepsilon) & 0 & 0 \\
0 & DG_{11} & DG_{12} \\
0 & DG_{21} & DG_{22}
\end{bmatrix},
\tag{77}
$$

where DG_{ij} stands for the 2×2 matrix

$$
DG =
\begin{bmatrix}
\frac{2}{c^r + d^r}((d^r - c^r)\lambda_1^r \mp 4c^r \rho_\varepsilon \cos(\varphi_\varepsilon)) & \pm 2\rho_\varepsilon \sin(\varphi_\varepsilon) \\
\frac{-2(c^i - d^i)}{c^r + d^r}(\lambda_1^r \pm \rho_\varepsilon \cos(\varphi_\varepsilon)) \mp 2\rho_\varepsilon \sin(\varphi_\varepsilon) & \mp 2\rho_\varepsilon \cos(\varphi_\varepsilon)
\end{bmatrix}.
\tag{78}
$$

According to the experimentally observed dynamics, states close to zero evolve toward a standing wave. Therefore, the zero will be unstable in the $\rho_1 = \rho_2 = \rho$ direction, and that will occur if $-2\lambda_1^r \mp 2\rho_\varepsilon \cos(\varphi_\varepsilon) > 0$. As A^2 must be a positive quantity, according to (76) $c^r + d^r < 0$. If the eigenvalues of DG have negative real parts, the fixed points will be stable in the α and δ directions also. The averaged intensity of these states can be computed as $< EE^* >$, with E given by

$$
E = e^{i(\varphi_1 - \frac{\delta}{2})}\rho(e^{i(\theta - \frac{\delta}{2})} + e^{-i(\theta - \frac{\delta}{2})})f(r)e^{i\omega_1 t}.
\tag{79}
$$

Therefore, the pattern corresponding to this state will look like a set of two bright spots either parallel ($\delta = 0$) or perpendicular ($\delta = \pi$) to the privileged axis as shown in Fig. 27.

Now let us state the conditions for switching to the central mode in such a way that the dynamics of our model closely resemble the experimental dynamics. Basically, we want the standing wave to be unstable in the ρ_1 and ρ_2 directions, so that initial states close to the standing wave, and with small components of ρ_0, evolve toward a central mode, as shown in Fig. 28.

In the $\rho_1 = \rho_2$, $\delta = 0$ subspace, the dynamics are given by

$$
\begin{aligned}
\dot{\rho}_0 &= (\lambda_0^r + a^r A^2 + b^r \rho_0^2)\rho_0, \\
\dot{A} &= (\lambda_1^r + \rho_\varepsilon \cos(\varphi_\varepsilon) + \frac{1}{2}(c^r + d^r)A^2 + e^r \rho_0^2)A.
\end{aligned}
\tag{80}
$$

In the (ρ_0, A) plane of the phase space, there are three fixed points O, SW, and C, with respective coordinates $(0,0)$, $(0, \sqrt{-2(\lambda_1^r + \rho_\varepsilon \cos(\varphi_\varepsilon))/(c^r + d^r)})$, and $(\sqrt{-\lambda_0^r/b^r}, 0)$. If the following happens,

$$
\begin{aligned}
\lambda_0^r - 2\frac{a^r}{c^r + d^r}(\lambda_1^r + \rho_\varepsilon \cos(\varphi_\varepsilon)) &> 0, \\
-\frac{e^r}{b^r}\lambda_0^r + (\lambda_1^r + \rho_\varepsilon \cos(\varphi_\varepsilon)) &< 0,
\end{aligned}
\tag{81}
$$

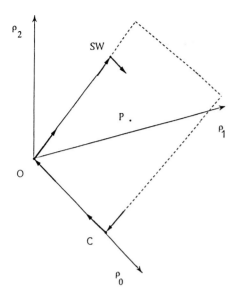

Fig. 28. Three-dimensional (ρ_0, ρ_1, ρ_2) projection of the phase space. The arrows indicate the local stability properties of the zero (O), central (C), and standing wave (SW) solutions, which lie on the (ρ_0, A) plane (dashed lines).

the fixed point SW is unstable in the ρ_0 direction and C is stable in the A direction. As $b^r > 0$, C is unstable in the ρ_0 direction. Notice that $D \equiv a^r e^r - b^r(c^r + d^r)/2$ must be negative for both conditions to be satisfied at the same time. If this happens, the (ρ_0, A) plane includes a fourth fixed point P with

$$\rho_0^2 = \frac{1}{D}\left(\frac{c^r + d^r}{2}\lambda_0^r - a^r(\lambda_1^r + \rho_\varepsilon \cos(\varphi_\varepsilon))\right),$$

$$A^2 = \frac{1}{D}\left(-e^r \lambda_0^r + b^r(\lambda_1^r + \rho_\varepsilon \cos(\varphi_\varepsilon))\right). \tag{82}$$

P undergoes a Hopf bifurcation if

$$\rho_0^2 = -\left(\frac{c^r + d^r}{2b^r}\right)A^2. \tag{83}$$

As a periodic orbit emerging from P gets close to the other fixed points, there will be a critical slowing down that will give rise to the PA phenomenon among the O, SW, and C states.

The numerical solutions of the equations (Fig. 29) show a qualitative agreement with the above considerations. The parameter values chosen for the simulations are reported in the figure captions. Notice that in the case of Fig. 29a, $(b^r/e^r) > (2a^r/(c^r + d^r))$, thus the eigenvalues of P in the $\alpha = \pi/2$, $\delta = 0$ plane are $\pm i$. For any initial condition, there is a periodic solution passing through it, suggesting the existence of integrals of motion. Notice that there is a heteroclinic solution connecting the three fixed points on the axes, and that the period of a periodic solution becomes larger the closer it is to the heteroclinic solution. Furthermore, the density of points along the trajectories shows a long persistence close to C, SW, and O and fast transitions in between, according to the experimental data of Fig. 4b in [Ar4].

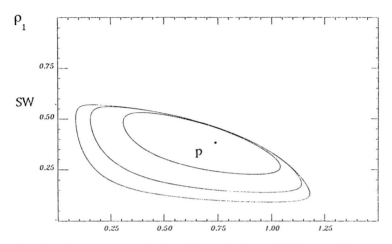

Fig. 29. The (ρ_0, ρ_1) projection of the solutions of (3) for suitable control parameters. P is marginally stable, and for any initial condition a periodic orbit is found. The closer a periodic orbit is to the three fixed points, the larger the period.

In the experiment of [Ar4], a strong pulling action due to the narrow frequency width of the photorefractive medium provides equal dressed frequencies for all transverse modes, even though the bare frequencies were different. Now, (72) refers to the modes dressed by their interaction with the medium, hence a frequency degeneracy $\omega_1 = \omega_0$ should be included. A resonance between the states with angular momenta ± 1 and the central state implies an additional symmetry. Namely, the time symmetry becomes [DnK]

$$B : (z_1, z_2, z_0) \rightarrow (e^{i\beta} z_1, e^{i\beta} z_2, e^{i\beta} z_0).$$

This allows additional terms to survive in the normal form equation and, as a result, the PA becomes CA as shown in detail in [GSS].

References

[ArH]   Instabilities and Chaos in Quantum Optics. Eds. Arecchi, F. T., Harrison, R. G., Springer, Berlin, 1987.

[NcP]   Nicolis, G., Prigogine, I.: Self Organization in Non-Equilibrium Systems. Wiley, New York, 1977.

[Hkn]   Haken, H.: Synergetics: An Introduction. Springer, Berlin, 1977; Advanced Synergetics. Springer, Berlin, 1983.
[Kr1]   Kuramoto, Y.: Chemical Oscillations, Waves and Turbulence. Springer, Berlin, 1984.
[VdP]   Nonequilibrium Dynamics in Chemical Systems. Eds. Vidal C., Pacult, A., Springer, Berlin, 1984.
[BGN]   Spatio Temporal Coherence and Chaos in Physical Systems. Eds. Bishop, A.R., Gruner, G., Nicolaenko, B., Physica D **23** 1986.
[Trn]   Turing, A.M.: Phil. Trans. R. Soc., London B **237** 1952, 37.
[ZhR]   Zhabotinsky, A.M., Rovinsky, A.B.: J. Stat. Phys. **48** 1987, 959.
[Vst]   Vastano, J.A., Pearson, J.E., Horsthemke, W., Swinney, H.L.: Phys. Lett. A **124** 1987, 320; J. Chem. Phys. **88** 1988, 6175.
[DlB]   Dewel, G., Borkmans, P.: Phys. Lett. A **84** 1987, 1335.
[LnG]   Landau, L.D., Ginzburg, V.L.: On the Theory of Superconductivity. In Collected Papers of L.D. Landau. Ed. Terhaar, D., Pergamon, Oxford, 1965, 217.
[Kr2]   Kuramoto, Y.: Prog. Theor. Phys. **71** 1984, 1182. In Chemical Oscillations, Waves and Turbulence. Ed. Haken, H., Springer, Berlin, 1984.
[Svs]   Sivashinsky, G.I.: Acta Astronaut. **4** 1977, 1177.
[PPP]   Pomeau, Y., Pumir, A., Pelce, P.: J. Stat. Phys. **37** 1984, 39.
[GGR]   Gaponov-Grekhov, A.V., Rabinovitch, M.I.: Usp. Phys. Nauk. **152** 1987, 159; Sov. Phys. Usp. **30** 1987, 433.
[CGL]   Coullet, P., Gil, L., Lega, J.: Phys. Rev. Lett. **62** 1989, 1619.
[Grn]   Goren, G., Procaccia, I., Rasenat, S., Steinberg, V.: Phys. Rev. Lett. **63** 1989, 1237.
[RcT]   Rica, S., Tirapegui, E.: Phys. Rev. Lett. **64** 1990, 878.
[BPJ]   Bohr, T., Pedersen, A.W., Jensen, M.H.: Phys. Rev. A **42** 1990, 3626.
[BLN]   Brand, H.R., Lomdahl, P.S., Newell, A.C.: Physica D **23** 1986, 345.
[Cst]   Castets, V., Dulos, E., Boissonade, J., De Kepper, P.: Phys. Rev. Lett. **64** 1990, 2953.
[KrO]   Kerner, B.S., Osipov, V.V.: Sov. Phys. Usp. **33** 1990, 679.
[Mnh]   Meinhardt, H.: Models of Biological Pattern Formation. Academic Press, New York, 1982.
[Bls]   Belousov, B.P.: Ref. Radiats, Med. 1958; Medgiz, Moscow, 1959, 145.
[MPH]   Muller, S.C., Plesser, T., Hess, B.: Science **230** 1985, 661; J. Stat. Phys. **48** 1987, 991.
[Arg]   Argoul, F., Arneodo, A., Richetti, P., Roux, J.C.: J. Chem. Phys. **86** 1987, 3325.
[TsK]   Tyson, J.J., Keener, J.P.: Physica D **32** 1988, 327.
[Mln]   Moloney, J.V., Hopf, F.A., Gibbs, H.M.: Phys. Rev. A **25** 1982, 3442; Laughlin, D.W., Moloney, J.V., Newell, A.C.: Phys. Rev. Lett. **51** 1983, 75.
[FrW]   Firth, W.J., Wright, E.M.: Phys. Lett. A **92** 1982, 211.
[Grn]   Green, C., Mindlin, G.B., D'Angelo, E.J., Solari, H.G., Tredicce, J.R.: Phys. Rev. Lett. **65** 1990, 3124.
[Brr]   Berry, M. In Physics of Defects. Ed. R. Balian et al., North Holland, Amsterdam, 1981, 456.
[VnN]   Von Neumann, J.: Theory of Self-Reproducing Automata. Ed. Burks, A. W., Univ. Illinois Press, 1966.
[Ar1]   Arecchi, F.T.: In Quantum, Optics. Ed. Glauber, R.J., Academic Press, New York, 1969, 57; Arecchi, F.T.: In Order and Fluctuations in Equilibrium and

Nonequilibrium Statistical Mechanics, Proc. XVII Int. Solvay Conf. on Physics. Eds. Nicolis, G., Dewel, G., and Turner, J.V., Wiley, New York, 1981, 107.

[Slv]   Proc. XVII Int. Solvay Conf. on Physics. Eds. Nicolis, G., Dewel, G., and Turner, J.V., Wiley, New York, 1981.

[And]   Anderson, P.W.: Science **177** 1972, 393.

[ScT]   Schawlow, A.L., Townes, C.H.: Phys. Rev. **112** 1958, 1940.

[VnL]   See the reference to the work of Von Laue, M., in A. Sommerfeld, Optics. Academic Press, New York, 1954, 194.

[Ar2]   Arecchi, F.T.: Phys. Rev. Lett. **24** 1965, 912.

[Ar3]   For a review, see Arecchi, F.T.: In Instabilities and Chaos in Quantum Optics. Eds. Arecchi, F.T., Harrison, R.G., Springer, Berlin, 1987, 9.

[Drn]   Doering, C.R., Biggon, J.D., Holm, D., Nicolaenko, B.: Nonlinearity **1** 1988, 279.

[CFT]   Constantin, P., Foias, C., Teman, R.: Physica D **30** 1988, 284.

[Shr]   Shraiman, B.I.: Phys. Rev. Lett. **57** 1987, 325.

[HhS]   Hohenberg, P.C., Shraiman, B.I.: Physica D **37** 1989, 109.

[Lmb]   Lamb, W.E.: Phys. Rev. A **134** 1964, 1429.

[Ar4]   Arecchi, F.T., Giacomelli, G., Ramazza, P.L., Residori, S.: Phys. Rev. Lett. **65** 1990, 2531.

[IMO]   Ikeda, K., Matsumoto, K., Otsuka, K.: Progr. Theor. Phys., Suppl. **99** 1989, 295.

[Ots]   Otsuka, K.: Phys. Rev. Lett. 199 **65**, 329.

[Knk]   Kaneko, K.: Physica D **54** 1991, 5.

[Tsd]   Tsuda, I.: Neural Networks **5** 1992, 313.

[BrZ]   Baranova, N.B., Zel'dovich, B.Ya.: Sov. Phys. JETP **53** 1981, 925; Baranova, N.B., Mamaev, A.V., Pilipetsky, N.F., Shkunov, V.V., Zel'dovich, B.Ya.: J. Opt. Soc. Am. **73** 1983, 525.

[BrW]   Born, M., Wolf, E.: Principles of Optics. North Holland Publishing Company, Amsterdam, 1962.

[Cll]   Coullet, P., Elphick, C., Gil, L., Lega, J.: Phys. Rev. Lett. **59** 1987, 884.

[Clb]   Ciliberto, S., Coullet, P., Lega, J., Pampaloni, E., Perez-Garcia, C.: Phys. Rev. Lett. **65** 1990, 2370.

[Ar5]   Arecchi, F.T., Giacomelli, G., Ramazza, P.L., Residori, S.: Phys. Rev. Lett., **67** 1991, 3749.

[AhB]   Ahlers, G., Behringer, R.P.: Prog. Theor. Phys. Suppl. **64** 1979, 186; Pocheau, A., Croquette, V., Le Gal, P.: Phys. Rev. Lett. **65** 1985, 1094.

[RbJ]   Ribotta, R., Joets, A.: In Cellular Structures and Instabilities. Ed. Wesfried, J.E., and Zaleski, S., Springer, Berlin, 1984; Rehberg, I., Rasenat, S., Steinberg, V.: Phys. Rev. Lett. **62** 1989, 756; Goren, G., Procaccia, I., Rasenat, S., Steinberg, V.: Phys. Rev. Lett. **63** 1989, 1237.

[GlR]   Gollub, J.P., Ramshankar, R.: In New Perspectives in Turbulence. Ed. Orszag, S., and Sirovich, L., Springer, Berlin, 1990.

[Kws]   Kawasaki, K.: Prog. Theor. Phys. Suppl. **79** 1984, 161; Coullet, P., Gil, L., Lega, J.: Phys. Rev. Lett. **62** 1989, 1619; Bodenschatz, E., Pesch, W., Kramer, L.: Physica D **32** 1988, 135.

[CGR]   Coullet, P., Gil, L., Rocca, F.: Opt. Commun. **73** 1989, 403.

[TIK]   Takeda, M., Ina, M., Kobayashi, S.: J. Opt. Soc. Am. **72** 1982, 156.

[Brm]   Brambilla, M., Battipede, F., Lugiato, L.A., Penna, V., Prati, F., Tamm, C., Weiss, C.O.: Phys. Rev. A **43** 1991, 5090.

[Sgm]   Siegman, A.E.: Lasers. University Sci. Books, San Francisco, 1987.

7 Pattern and Vortex Dynamics in Photorefractive Oscillators   215

[GLM] Gil, L., Lega, J., Meunier, J.L.: Phys. Rev. A **41** 1990, 1138.
[GPR] Goren, G., Procaccia, I., Rasenat, S., Steinberg, V.: Phys. Rev. Lett. **63** 1989, 1237.
[Ar6] Arecchi, F.T., Boccaletti, S., Ramazza, P.L., Residori, S.: Transition from Boundary- to Bulk-Controlled Regimes in Optical Pattern Formation. Submitted for publication.
[Mnn] Manneville, P.: In Dissipative Structures and Weak Turbulence, Academic Press, Inc., San Diego, CA, 1990.
[EKR] Ezerskii, A.B., Korotin, P.I., Rabinovich, M.I.: JETP Lett. **41** 1985, 157; Tufillaro, N., Ramshankar, R., Gollub, J.P.: Phys. Rev. Lett. **62** 1989, 422.
[FvK] Fauve, S., Kumar, K., Laroche, C., Beysens, D., Garrabos, Y. Phys. Rev. Lett. **68** 1992, 3160.
[LgL] Lugiato, L.A., Lefever, L.: Phys. Rev. Lett. **58** 1987, 2209.
[GLV] Grynberg, G., Le Bihan, E., Verkerk, P., Simoneau, P., Leite, J.R.R., Bloch, D., Le Boiteux, S., Ducloy, M.: Opt. Comm. **67** 1988, 363; Petrossian, A., Pinard, M., Maitre, A., Courtois, J.-Y., Grynberg, G.: Europhys. Lett. **18** 1992, 689.
[McE] Macdonald, R., Eichler, H.J.: Opt. Comm. **89** 1992, 289.
[Kgl] Kogelnik, H.: In Lasers: A Series of Advances. Ed. Levine, A.K., Dekker, New York, 1966.
[Cm1] Even though unnecessary for the purposes of this paper, we recall that the decorrelation mechanism in a photorefractive crystal is due to the diffusion of the space-charge wave, which provides the refractive index grating. The decorrelation length measured here coincides with the diffusion length, as discussed in detail in Arecchi, F.T., Boccaletti, S., Giacomelli, G., Puccioni, G.P., Ramazza P.L., Residori, S.: Physica D **61** 1992.
[GnH] Günter, P., Huignard, J.P.: Photorefractive Materials and Their Applications, Springer-Verlag, Berlin, 1988.
[DAF] D'Alessandro, G., Firth, W.J.: Phys. Rev. A **46** 1992, 537.
[Ar7] Arecchi, F.T., Boccaletti, S., Mindlin, G.B., Perez-Garcia, C.: Phys. Rev. Lett. **69** 1992, 3723.
[Ar8] Arecchi, F.T., Boccaletti, S., Puccioni, J.P.: Pattern Formation and Competition in Nonlinear Optical Oscillators. Submitted for publication.
[Kkh] Kukhtarev, N.V., Markov, V.B., Odulov, S.G., Soskin, M.S., Vinetskii, V.L.: Ferroelectrics **22** 1979, 949.
[Fnb] Feinberg, J., Helman, D., Tanguay, Jr., A.B., Hellwarth, R.W.: J. Appl. Phys. **51** 1980, 1279.
[JHF] Jaura, R., Hall, T.P., Foote, P.D.: Opt. Eng. **25** 1986, 1068.
[Krn] Krainak, M.A.: Abstract presented to the topical meeting "PhotoreFractive Materials, Effects and Devices II," Aussois, France, 1990.
[DAl] D'Alessandro, G.: Phys. Rev. A **46** 1992, 2791.
[ACA] Arizmendi, L., Cabrera, J.M., Agullo-Lopez, F.: Int. J. Opt. **7** 1992, 149.
[Rmz] Ramazza, P.L.: PhD Thesis, 1989, unpublished
[Mss] Messiah, A.: Quantum Mechanics, North Holland, Amsterdam, 1962.
[GSS] Golubitsky, M., Stewart, I., Schaeffer, D.: Singularities and Groups in Bifurcation Theory. Appl. Math. Sciences **69** Springer-Verlag, New York, 1983.

[DnK] Dangelmayr, G., Knobloch, E.: Nonlinearity **4** 1991, 399.

[Cm2] Both the previous reference, as well as a recent application of the symmetry
       arguments to the patterns of a CO_2 laser by D'Angelo, E.J., Izaguirre, E.,
       Mindlin, G.B., Huyet, G., Gil, L., Tredicce, J.R.: Phys. Rev. Lett. **68** 1992,
       3702, refer to a two-mode interaction and are concerned with the steady solu-
       tions, whereas here we study the time-dependent trajectories of a three-mode
       case.

8 From Hamiltonian Mechanics to Continuous Media. Dissipative Structures. Criteria of Self-Organization

Yu. L. Klimontovich

Department of Physics, Moscow State University, 117234 Moscow, Russia

Introduction

This paper presents some main ideas and the results of modern statistical theory in regard to macroscopic open systems.

We begin from the demonstration of the necessity for and possibility of developing a unified description of kinetic, hydrodynamic, and diffusion processes in nonlinear macroscopic open systems based on a generalized kinetic equation.

A derivation of the generalized kinetic equation is based on the concrete physical definition of continuous media. A "point" of a continuous medium is determined by definition of physically infinitesimal scales. On the same basis, the definition of the Gibbs ensemble is given. The Boltzmann gas and a fully ionized plasma are used as the test systems.

For the transition from the reversible Hamilton equations to the generalized kinetic equations, the dynamic instability of the motion of particles plays a constructive role.

The kinetic equation for the Boltzmann gas consists of two dissipative terms:

1. The "collision integral," which is defined by processes in a velocity space
2. The additional dissipative term of the diffusion type in the coordinate space

Owing to the latter, the unified description of the kinetic, hydrodynamic, and diffusion processes for all values of the Knudsen number becomes possible.

The H-theorem for the generalized kinetic equation is proved. An entropy production is defined by the sum of two independent positive terms corresponding to redistribution of the particles in velocity and coordinate space, respectively.

An entropy flux also consists of two parts. One is proportional to the entropy and the other is proportional to the gradient of entropy. The existence of the second term follows to give the general definition of the heat flux for any values of the Knudsen number, which is proportional to the gradient of entropy. This general definition for small Knudsen numbers and a constant pressure leads to the Fourier law.

The equations of gas dynamic for any class of distribution functions follow from the generalized kinetic equation without the perturbation theory for the Knudsen number. These equations differ from the traditional ones by taking into account the self-diffusion processes.

The generalized kinetic equation for description of the Brownian motion and autowave processes in active media is considered. The connection with reaction diffusion equations – the Fisher-Kolmogorov-Petrovskii-Piskunov (FKPP) and the Ginzburg-Landau equations – are established.

We discuss the connection between the diffusion of particles in a restricted system with the natural flicker $(1/f)$ noise in passive and active systems.

The criteria of the relative degree of order of the states of open systems – the criteria of self-organization – are presented.

1 The Transition from Reversible Equations of Mechanics to Irreversible Equations of Statistical Theory

1.1 Physical Definition of Continuous Medium

To describe the transition to irreversible equations of the statistical theory, it is necessary to take into account the structure of the "continuous medium." In other words, it is necessary to give the concrete definition of the infinitesimal scales – to give the physical definition of the notion "point" [K1, K2, K3, K4, K5, K6].

Obviously, it is impossible to give a unified definition of the physically infinitesimal scales for all systems. We shall introduce these scales for two simple cases: the Boltzmann gas and a fully ionized plasma. Such definitions are different for the kinetic and hydrodynamic region of scales.

The Kinetic Region. A rarefied (Boltzmann) gas and a rarefied, fully ionized plasma, respectively, are characterized by the dimensionless small density and plasma parameters

$$\epsilon = nr_o^3; \qquad \mu = 1/nr_D^3, \tag{1}$$

where r_o is the diameter of atoms, r_D is the Debye radius, and n is the mean density of the number of particles. These parameters determine the connection between the infinitesimal scales τ_{ph}, l_{ph} and the corresponding "collision" parameters τ, l of the Boltzmann gas and the Debye plasma.

We denote by $N_{ph} = nV_{ph}$ the number of particles in the physically infinitesimal volume and define the physically infinitesimal time as

$$\tau/N_{ph} = \tau_{ph}, \quad N_{ph} = nV_{ph} \sim nl_{ph}^3, \quad \tau_{ph} = l_{ph}/v_T. \tag{2}$$

Thus defined, τ_{ph} is the time within which any one particle out of the number N_{ph} in the volume V_{ph} undergoes a collision.

The physically infinitesimal scales for the Boltzmann gas follow from the expressions (2)

$$\tau_{ph} \sim \sqrt{\epsilon}\tau \ll \tau, \quad l_{ph} \sim \sqrt{\epsilon}l \ll l, \quad N_{ph} \sim \frac{1}{\sqrt{\epsilon}} \gg 1. \tag{3}$$

The corresponding kinetic characteristic for a rarefied plasma is $(a = e, i)$

$$\tau_{ph}^{(a)} \sim \mu\tau^{(a)} \ll \tau^{(a)}, \quad l_{ph}^{(a)} \sim r_D \sim \mu l^{(a)} \ll l^{(a)}, \quad N_{ph} \sim 1/\mu \gg 1. \tag{4}$$

We see that for the Boltzmann gas and a rarefied plasma the general conditions imposed on the physically infinitesimal scales are satisfied.

The Gas Dynamic Region. In this case, the relaxation time is defined by the characteristic scale of the system $L : \tau_D = L^2/D$. The role D is played by viscosity ν, thermal conductivity χ, and the self-diffusion coefficient D. Here $D = \nu = \chi$. The physically infinitesimal characteristics are defined now by relations

$$\tau_{ph}^G \sim \frac{\tau_D}{N_{ph}^G}, \quad l_{ph}^G \sim \frac{L}{N^{1/5}}, \quad N_{ph}^G \sim N^{2/5}, \tag{5}$$

here $N = nL^3$.

The definition of physically infinitesimal scales allows us to use the notion of the "continuous medium" for the kinetic and hydrodynamic description, respectively. To illustrate the importance of this notion, we consider the following example.

The Maximum of the Reynolds Number in the Kolmogorov Theory. In the Kolmogorov theory of a developed turbulence, the number of collective (turbulent) degrees of freedom $N_{turb} = (L/L_o)^3$ is connected with the Reynolds number by the relation [K4, MY, LL]

$$N_{turb} \sim Re^{9/4}. \tag{6}$$

Here L and L_o, respectively, are the largest and the smallest scales of the developed turbulent motion.

From the definition N_{turb}, it follows that the maximum value of the Reynolds number in the Kolmogorov theory is restricted by the condition $L_o > l_{ph}^G$ and, as a result,

$$(Re)_{\max} \sim N^{4/15}; \quad N = nL^3. \tag{7}$$

We see that, as a consequence, the limit $Re \Rightarrow \infty$, which is used frequently in the theory of turbulent motion for the "physical continuous medium," cannot be realized.

1.2 The Gibbs Ensemble for Nonequilibrium Processes

The statistical ensemble was introduced to describe the equilibrium state. In this case, the number of controllable degrees of freedom is very small, hence, the indeterminacy in the microscopic states of the system is extremely high. This way of defining the statistical ensemble of arbitrary nonequilibrium states is not suitable. What then can be done?

Assume that we have chosen the concrete definition of the physically infinitesimal scales. We can then assume that the indeterminacy of defining the microscopic states of systems in the Gibbs ensemble is governed by the indeterminacy in the states of particles confined within the volume V_{ph}. This allows us to carry out the operation of averaging, or smoothing, dynamic distributions over a physically infinitesimal volume.

1.3 The Unified Definition of "Continuous Medium" Averaging Over Physically Infinitesimal Volume [K10, K11]

To develop the unified description of the kinetic and hydrodynamic processes, it is necessary to use the equation $(V_{ph}^G)_{min} = V_{ph}$ to express the minimal physically infinitesimal volume $(V_{ph}^G)_{min} \equiv L_{min}^3$ via small density parameter ϵ.

$$L_{min} \sim (N_{ph})^{1/2} l_{ph} \sim l/(N_{ph})^{1/2}, \quad N_{min} = n L_{min}^3 \sim \epsilon^{-5/4}. \qquad (8)$$

Thus, for example, at atmospheric pressure the density parameter $\epsilon \approx 10^{-4}$ and, therefore, the number of particles in the "point" N_{min} is in the order of 10^5.

Let us consider a local random function $N(r, p, t)$ – the microscopic phase density in the six-dimensional space of coordinate and momenta

$$N(r, p, t) = \sum_{1 \leq i \leq N} \delta(r - r_i(t)) \, \delta(p - p_i(t)) \qquad (9)$$

and introduce the new function smoothed over the distribution

$$F(\rho) = (2\pi L_{min}^2)^{-3/2} \exp(-\rho^2/(L_{min})^2),$$
$$\tilde{N}(r, p, t) = \int N(r - \rho, p, t) F(\rho) d\rho. \qquad (10)$$

Initially, we shall use the following dynamical reversible equations for the microscopical function (9) [K1, K2, K3, K4, K5, K6, K7]

$$\frac{\partial N}{\partial t} + v \frac{\partial N}{\partial r} + F^m(r, t) \frac{\partial N}{\partial p} = 0,$$
$$F^m(r, t) = F_o - \frac{\partial}{\partial r} \int \Phi(|r - r'|) N(r', p', t) dr' dp'. \qquad (11)$$

Here F^m is the microscopic force.

To obtain the generalized kinetic equation on the basis of this dynamic reversible equation, it is necessary to take into account the dynamic instability of the motion of atoms.

1.4 The Constructive Role of the Dynamic Instability of the Motion of Atoms

The kinetic theory of gases is traditionally constructed (the Bogolubov-Born-Green-Kirkwood-Yvon theory, for example) without using the principal ideas of the dynamic theory – the concepts of dynamic instability, dynamic chaos, K-entropy, and mixing. Let us show that these enable us to elucidate more fully the causes of irreversibility and to obtain the generalized kinetic equation.

The first steps in this direction were made by N.S. Krylov in 1950 [Kr], when he dealt with the substantiation of statistical physics. The importance of the dynamic instability of motion in justifying the irreversible equations of statistical physics is discussed elsewhere in Prigogine [Pr, PrS] and [RSC, K8].

Let us make some elementary quantitative estimates that will connect the time interval τ_{ph}, introduced here by way of the physically infinitesimal time scale, with the minimum time of development of dynamic instability .

The motion of atom spheres in a Boltzmann gas, like the motion of balls in Sinai billiards, is dynamically unstable. For a single atom, the time of development of instability – or, in other words, the characteristic time of exponential divergence of initially close trajectories of two atoms – is of the order of the free path time [RSC, K8]

$$\tau_{inst} \sim \tau. \tag{12}$$

This estimate relates to the path of a single chosen atom. We take into account that in the process of smoothing over the physically infinitesimal volume V_{ph}, all N_{ph} particles contained therein are indistinguishable. We can, therefore, introduce, along with τ_{inst}, the characteristic time of development of instability for any one particle within V_{ph}. Let us denote the new quantity by $(\tau_{inst})_{ph}$. This time is N_{ph} times shorter than τ_{inst}, and hence, given the definition (12), we come to the following estimate for the minimum characteristic time of development of dynamic instability in the motion of atoms in a Boltzmann gas:

$$(\tau_{inst})_{ph} \sim \tau_{ph}. \tag{13}$$

Thus, the characteristic time for the development of instability of motion per particle within N_{ph} is of the order of the physically infinitesimal time scale τ_{ph}. This is another argument in favor of our method of defining τ_{ph}.

The dynamic instability of the motion of atoms in Boltzmann gas leads to mixing and, therefore, facilitates the transition from the reversible Hamiltonian equations to the much simpler Boltzmann kinetic equation. We see this as the constructive role of the dynamic instability of motion in the formulation of the statistical theory of nonequilibrium processes.

Macroscopic characteristics, for instance the distribution function, can be termed the *functions of order*, insofar as they single out and describe a more ordered motion against the background of molecular chaotic motion. In other words, the functions of order reveal the statistical laws.

This illustration, drawn for a Boltzmann gas, in no way exhausts the constructive role of dynamic instability.

In fact, in statistical theory the dynamic instability can be associated not only with the atoms, but also with the macroscopic characteristics of the system. The latter kind of instability was first discovered with the model equations of thermal convection in a fluid, based on the Lorenz equations [Lor]. Examples of physically feasible systems in which a large amount of feedback gives rise to dynamic instability of the macroscopic characteristics are lasers and electrical devices [Lor, H1, Ani, NeL, K5, K6].

In connection with the above view concerning the constructive role of dynamic instability of the motion of atoms in a Boltzmann gas, some questions inevitably arise. Can the dynamic instability of the motion of the macroscopic characteristics also play a constructive role? Will this kind of instability lead to more sophisticated dissipative structures or to "dynamic chaos"? In what relation does physical chaos stand to dynamic chaos?

To answer these questions, we need a criterion for the relative degree of order or organization (or alternatively, chaoticity) of the nonequilibrium states of open systems. We shall use the Boltzmann-Gibbs-Shannon entropy renormalized to a given mean effective energy – the effective Hamiltonian function of the open system.

The quantitative assessment of the relative order will be based on the S-Theorem. The S-Theorem will be used to check on the proper choice of controlling parameters, then to assess the relative degree of order. It is also possible to organize the optimization of the search for the most ordered states in the space of the controlling parameters of the open systems.

Basing our reasoning on the criterion of the relative degree of order, we shall use some examples to demonstrate that the processes of self-organization are also possible in the presence of dynamic instability of the motion of macroscopic characteristics of open systems. However, we shall first continue to discuss the question about the generalized kinetic equation.

2 The Unified Description of Kinetic and Hydrodynamic Motion

2.1 The Generalized Kinetic Equation [K5, K6, K10, K11]

To take into account the existence of the dynamic instability of the motion of atoms, we can introduce into the reversible dynamic equation (11) the term

$$-\frac{1}{\tau_{ph}}(N(r,p,t) - \tilde{N}(r,p,t)), \qquad (14)$$

which describes the relaxation from the dynamic microscopic phase density $N(r,p,t)$ to the smoothed distribution $\tilde{N}(r,p,t)$. After averaging over the Gibbs ensemble, we can obtain the equation for the distribution function $f(r,p,t) = <N(r,p,t)>/n$ with the additional relaxation term

$$\frac{\partial f}{\partial t} + v\frac{\partial f}{\partial r} + F(r,t)\frac{\partial f}{\partial p} = -\frac{1}{n}\frac{\partial <\delta F \delta N>}{\partial p} - \frac{1}{\tau_{ph}}(f(r,p,t) - \tilde{f}(r,p,t)),$$

$$F(r,t) = F_o - n \int \frac{\partial \Phi(|\ r - r'\ |)}{\partial r} f(r',p',t)dr'dp'. \qquad (15)$$

On the right side of this equation, there are now two dissipative terms. The first term describes the dissipation by processes in the velocity space – by "collisions" [K1, K2, K3, K4, K5, K6, LL]. This term can be presented as the Boltzmann collision integral for a rarefied gas or as the Landau or the Balescu-Lenard "collision integral" for fully ionized plasmas.

To obtain the generalized kinetic equation for the unified description of kinetic and hydrodynamic processes, it is natural to use the simplest form for this collision integral. Indeed, we saw the number of particles $N_{min} \sim \epsilon^{-5/4}$ and in the case of a rarefied gas much more unity ($N_{min} \approx 10^5$ at atmospheric pressure), therefore, it is possible to use not the Boltzmann form, but the nonlinear Fokker-Planck form for the collision integral

$$-\frac{1}{n}\frac{\partial <\delta F \delta N>}{\partial p} \equiv I_{(v)}(r,p,t) = \frac{\partial}{\partial v}\left[D_{(v)}(r,t)\frac{\partial f}{\partial v}\right] + \frac{\partial}{\partial v}\left[\frac{1}{\tau}(v-u(r,t))f\right]. \qquad (16)$$

The diffusion coefficient in the velocity space is defined by the local temperature

$$D_{(v)}(r,t) = \frac{1}{\tau}\ae T(r,t) = \frac{m}{3}\int (v - u(r,t))^2 f(r,p,t)dv, \qquad (17)$$

where $u(r,t)$ is a local hydrodynamic velocity and τ is a "collision" time.

To obtain the final form for the second collision integral, we can use the expansion on the "physical Knudsen number"

$$(Kn)_{ph} = \frac{l_{ph}}{L}. \quad \text{Here } l_{ph} = L_{min}. \qquad (18)$$

If the mean force $F(r,t)$ is not zero, then it is necessary to change the definition (10) of the smoothing function: the mean value $< \rho >= (b/m)F\tau_{ph}$. Here b is the mobility of atoms.

By expansion of the additional dissipative term in the equation (15) on the "physical Knudsen parameter" (18), we have the following expression for the new collision integral:

$$-\frac{1}{\tau_{ph}}(f - \tilde{f}) \equiv I_r(r,p,t) = \frac{\partial}{\partial r}\left(\frac{L_{min}^2}{\tau_{ph}}\frac{\partial f}{\partial r}\right) - \frac{\partial}{\partial r}\left(\frac{b}{m}Ff\right). \qquad (19)$$

To obtain the final form for the new collision integral, we must take into account the definitions (3) and (8) for the scales τ_{ph}, L_{min}. As a result, we have

$$\frac{L_{min}^2}{\tau_{ph}} \approx \frac{l^2}{\tau} = D = b\frac{\ae T}{m}, \quad b = \tau. \qquad (20)$$

Here D is the space self-diffusion coefficient, which is connected with the mobility coefficient by the Einstein relation. For nonequilibrium states, the Einstein relation connects local characteristics, therefore, the final form for the new collision integral is

$$I_r(r,p,t) = \frac{\partial}{\partial r}\left[D(r,t)\frac{\partial f}{\partial r} - \frac{b}{m}Ff\right], \quad D(r,t) = b\frac{\text{æ}T}{m}. \tag{21}$$

It is useful to remind that D is one of the three kinetic coefficients: kinematic viscosity ν, temperature conductivity χ, and self-diffusion D. For our model, these three coefficients are equal, $D = \nu = \chi$. When there is a distinction between D, ν, and χ, it is possible to take these distinctions into account in the hydrodynamics equation.

The inclusion of self-diffusion into the equations of hydrodynamics has been suggested more than once (see [K5, K6]). It was argued that in the first self-diffusion distorts the conventional structure of the equations of hydrodynamics; and in the second, from the kinetic Boltzmann equation, it follows that self-diffusion is absent from the equations of hydrodynamics, because the transfer of matter is determined completely by the convective flow ρu. Landau and Lifshitz [LL] (p. 274 of the Russian edition) observed that the inclusion of self-diffusion into the hydrodynamic equations may result in the violation of the condition that entropy production in a closed system should be positive, thus violating the second law of thermodynamics.

Of course the last two arguments cannot be dismissed, and in the next sections we shall discuss them.

At this point, we note only the following:

In the equilibrium state, the kinetic equation (15) is satisfied by the Maxwell-Boltzmann distribution. The left-hand side, determined by the nondissipative terms, goes to zero independently of the dissipative terms on the right-hand side, each of which also goes to zero.

The first of the latter, which is either the Boltzmann collision integral or, more simply, the collision integral (16) through cancelling out the collisions of two types forms the Maxwell distribution.

The second dissipative term on the right-hand side also describes the balance of two dissipative flows. One of these flows is caused by the external force and is proportional to the mobility, while the second is the flow of matter due to self-diffusion. In this way, the existence of the equilibrium Boltzmann distribution can be regarded as a manifestation of self-diffusion.

3 The Equation of Entropy Balance. The Heat Flow

The local Boltzmann entropy is defined by the expression

$$S(r,t) \equiv \rho(r,t)s(r,t) = -\text{æ}n\int \ln(nf(r,p,t))f(r,p,t)dp. \tag{22}$$

At the condition $F = 0$, the equation of entropy balance for the generalized kinetic equation (15) can be presented in the following elegant form

$$\frac{\partial \rho s}{\partial t} + \frac{\partial}{\partial r}\left[\left(\rho u - D\frac{\partial \rho}{\partial r}\right)s\right] = \frac{\partial}{\partial r}\left(D\rho\frac{\partial s}{\partial r}\right) + \sigma(r,t). \tag{23}$$

The full entropy production is defined by the expression

$$\sigma(t) = \text{æ} n \int D_v f \left(\frac{\partial}{\partial v} \ln \frac{f}{f_{loc}} \right)^2 dr dp + \text{æ} n \int D f \left(\frac{\partial}{\partial r} \ln \frac{f}{f_o} \right)^2 dr dp \geq 0. \quad (24)$$

Here f_o is the Maxwell distribution and the f_{loc} is the local one.

We see that the entropy production is the sum of two positive terms, which are defined, respectively, by the changing of the distribution function in the velocity and coordinate spaces.

The entropy flow j_S consists of two parts

$$j_S(r,t) = j_{con} + j_{dif}. \quad (25)$$

The convective part, j_{con}, is proportional to the full flow of matter and the entropy $s(r,t)$, but the diffusion part, j_{dif}, is proportional to the gradient of entropy $s(r,t)$. It is natural to define the heat flow by the expression [K10, K11]

$$q(r,t) = T(r,t)j_{dif} = -T(r,t)D\rho\frac{\partial s}{\partial r}. \quad (26)$$

For the region of small values of the Knudsen number, when there is local equilibrium, the heat flow is defined by the gradients of density and temperature

$$q(r,t) = \frac{\text{æ}}{m} D \frac{\partial \rho}{\partial r} - c_v \rho \chi \frac{1}{T} \frac{\partial T}{\partial r}, \quad c_v = \frac{3}{2} \frac{\text{æ}}{m}. \quad (27)$$

Only for slow processes, when the pressure $p = const$ and the gradient of density is proportional to the temperature gradient, the heat flow is defined by the Fourier law

$$q(r,t) = -\lambda \frac{\partial T}{\partial r}, \quad \lambda = c_p \rho \chi, \quad c_p = \frac{5}{2} \frac{\text{æ}}{m}. \quad (28)$$

Here c_p is a heat capacity at constant pressure.

4 Equations of Hydrodynamics with Self-Diffusion

The representation of the collision integral $I_v(r,p,t)$ in the form (16) is based on the fact that the number of particles in a "point" is $N_{\min} \sim \epsilon^{-5/4} \gg 1$. For the same reason, it is possible to restrict the class of distribution functions $f(r,p,t)$ by condition

$$f(r,p,t) = f(r, m \mid v - u(r,t) \mid, t). \quad (29)$$

Then the closed system of equations for the gas-dynamical functions follows from the generalized kinetic equation (15) with collision integrals (16) and (21), without the perturbation theory for the Knudsen number.

To obtain the gas-dynamical equations for the functions $\rho(r,t)$, $u(r,t)$, and $T(r,t)$, we substitute distribution (29) into kinetic equation (15) and go over to equations in the relevant moments of the distribution (29). In the result, we come to the set of equations with the self-diffusion taken into account with the set of equations of hydrodynamics for self-diffusion

$$\frac{\partial \rho}{\partial t} + \frac{\partial}{\partial r} j(r,t) = 0, \quad J = \rho u - D \frac{\partial \rho}{\partial r} + \frac{b}{m} F \rho, \tag{30}$$

$$\frac{\partial \rho v_i}{\partial t} + \frac{\partial}{\partial r_j}(j_j u_i) = -\frac{\partial p}{\partial r_i} + \rho \nu \frac{\partial^2 u_i}{\partial r^2} + \frac{\rho}{m} F, \quad p = \frac{\rho}{m} \ae T, \tag{31}$$

$$\frac{\partial}{\partial t}\left(\frac{\rho u^2}{2} + \epsilon(r,t)\right) + \frac{\partial}{\partial r_i}\left[j_i\left(\frac{\rho u^2}{2} + \epsilon\right) + u_i p - \rho \nu u_j \frac{\partial u_i}{\partial r_j} - c_v \rho \chi \frac{\partial T}{\partial r_i}\right] = \frac{\rho}{m} F u,$$

$$\epsilon(r,t) = \frac{\rho}{m}\frac{3}{2}\ae T. \tag{32}$$

From the equation of continuity, it follows that the transfer of matter is now determined by three flows: convective transfer, self-diffusion, and the flow is defined by the mobility of atoms. If we take the external force into account, the equation takes the form

$$\frac{\partial \rho}{\partial t} + \frac{\partial}{\partial r}\left(\rho u - D\frac{\partial \rho}{\partial r} + \frac{b}{m} F \rho\right) = 0. \tag{33}$$

Equations (30) - (33) take into account the fact that the kinetic coefficients D, ν, χ may not be all the same. Observe also that in this approximation the tensor of viscous stress has a different form from the conventional representation. Namely, here $\pi_{ij} = -\eta \partial u_i / \partial r_j, \eta = \rho \nu$.

In the approximation of incompressible fluid, when $\rho = const$ and $F = const$, and the temperature (the variance of the velocity) is zero, the distribution function $f(r,p,t)$ takes the form

$$f(r,v,t) = \delta[v - u(r,t)], \tag{34}$$

and the equations of hydrodynamics coincide with the Navier-Stokes equations

$$\frac{\partial u}{\partial t} + (u \, grad)u = -\frac{1}{\rho} grad \, p + \nu \Delta u + \frac{1}{m} F,$$

$$div \, u = 0, \quad \Delta p = -\rho \frac{\partial u_i}{\partial r_j}\frac{\partial u_j}{\partial r_i}. \tag{35}$$

The equation for the pressure follows from the equation of continuity.

We can remark here that the Navier-Stokes equations, although extremely efficient, are intrinsically contradictory, because in this approximation the entropy in a closed system does not change and, at the same time, the production of entropy (24) is not zero.

For the class of distribution functions (29), to calculate the entropy production it is necessary to know the solution of the kinetic equation. Only for the local

Maxwell distribution, the entropy production is defined completely by functions $\rho(r,t)$, $u(r,t)$, and $T(r,t)$

$$\sigma(r,t) = \frac{\text{æ}}{m}\left[D\rho\left(\frac{grad\rho}{\rho}\right)^2 + \nu\rho\frac{m}{\text{æ}T}\left(\frac{\partial u_i}{\partial r_j}\right)^2 + \frac{3}{2}\chi\rho\left(\frac{gradT}{T}\right)^2\right] \geq 0. \tag{36}$$

Thus, the production of entropy is also non-negative when self-diffusion is taken into account, which is in full compliance with the second law of thermodynamics.

Observe finally that in the incompressible fluid approximation from (29) and (36), it follows that the entropy production is nonzero, while the entropy itself remains constant. Therefore, to describe thermal processes, we must go beyond the incompressible fluid approximation, for instance, to the Boussinesq approximation, to solve the Navier-Stokes equations together with the heat transfer equation

$$\frac{\partial T}{\partial t} + (vgrad)T = \chi\frac{\partial^2 T}{\partial r^2} + \frac{2}{3}\frac{m\nu}{\text{æ}}\left(\frac{\partial u_i}{\partial r_j}\right)^2. \tag{37}$$

We should like to underline that the Boussinesq approximation is valid *only* for the local Maxwell distribution. In more general cases, it is necessary to use the generalized kinetic equation.

5 Effect of Self-Diffusion on the Spectra of Hydrodynamic Fluctuations [K5, K6]

To calculate equilibrium hydrodynamic fluctuations, we use the linearized equations of hydrodynamics (30) - (32). In the linear approximation, the velocity field for the Fourier components is represented as a sum of the transverse (with respect to the wavevector $\mathbf{k}$) and the longitudinal parts. The inclusion of self-diffusion has no effect on the fluctuations of the transverse field of velocity, so the width of the spectrum of fluctuations is again determined by $\nu\mathbf{k}^2$, where ν is the kinematic viscosity.

The set of equations for the Fourier components of density, longitudinal velocity, temperature, and pressure, with due account for self-diffusion, now has the form

$$(-i\omega + Dk^2)\delta\rho + \rho i(k\delta u) = 0;$$

$$(-i\omega + \nu k^2)i(k\delta u)\rho = k^2\delta p;$$

$$-i\omega\delta p + pDk^2\frac{\delta\rho}{\rho} = \frac{p}{T}\chi k^2\delta T + \frac{5}{3}pi(k\delta u); \tag{38}$$

$$p = \frac{\rho}{m}\text{æ}T, \quad \delta p = \frac{\text{æ}}{m}\rho\delta T + \frac{\text{æ}T}{m}\delta\rho.$$

In calculations of fluctuations, two regions are usually distinguished

1. low frequencies, when (for $D = \nu = \chi$) $\omega, Dk^2 \ll kv_{sound}$
2. high frequencies, when $Dk^2 \ll \omega \sim kv_{sound}$

Calculations of equilibrium hydrodynamic fluctuations on the basis of hydrodynamics equations can be found elsewhere [LiP, K4]. Here we shall only indicate the basic changes in the spectra that occur when self-diffusion is taken into account.

For low frequencies, the spectra of fluctuations of density, temperature, and entropy are now given by

$$(\delta\rho\delta\rho)_{\omega,k} = \frac{2Dk^2}{\omega^2 + (Dk^2)^2}\frac{\ae\rho}{c_p}; \quad (\delta T\delta T)_{\omega,k} = \frac{2\chi k^2}{\omega^2 + (\chi k^2)^2}\frac{\ae T^2}{\rho c_p}; \quad (39)$$

$$(\delta S\delta S)_{\omega,k} = \left[\left(1 - \frac{c_v}{c_p}\right)\frac{2Dk^2}{\omega^2 + (Dk^2)^2} + \frac{c_v}{c_p}\frac{2\chi k^2}{\omega^2 + (\chi k^2)^2}\right]c_p\frac{\ae}{\rho}. \quad (40)$$

When self-diffusion is neglected, the widths of all these spectra are the same and are determined by the temperature conductivity χ. This results in a rigid correlation between the fluctuations of density and temperature at all temperatures $[\delta\rho(\omega,\mathbf{k})/\rho = \delta T(\omega,\mathbf{k})/T]$. When self-diffusion is included, this condition is removed and the width of the spectrum of density fluctuations is determined by the coefficient of self-diffusion; and the width of the spectrum of temperature fluctuations depends on the coefficient of temperature conduction χ.

Total correlation is only observed for characteristics that are integral with respect to ω. This correlation is dictated by the condition of local thermodynamic equilibrium.

The spectrum of fluctuations of the entropy is represented as a sum of two spectral lines, the relative contributions of which depend on the ratio between the heat capacities c_v and c_p.

Self-diffusion also affects the spectra of fluctuations at high frequencies. The line width is now determined by the combination of the three dissipative characteristics D, ν, χ

$$\gamma = \frac{1}{2}\left[\nu + \frac{c_v}{c_p}D + \left[1 - \frac{c_v}{c_p}\right]\chi\right]k^2. \quad (41)$$

Thus, the old problem – whether to include the contribution of self-diffusion into the equations of hydrodynamics – can also be resolved by analyzing experimental data on the spectra of molecular scattering in liquids.

The kinetic approach to the description of hydrodynamic motion described here can be extended to the case of turbulent motion.

6 The Kinetic Approach in the Theory of Self-Organization – Synergetics. Basic Mathematical Models

Both van der Pol generators and the more complicated oscillators can serve as elements of active (excitable) media. The first mathematical models of active media were proposed four decades ago in the well-known works of Wiener and Rosenbluth and Gelfand and Tsetlin (see in [RSC]). Landau's work (1944) on the origins of turbulence also made a fundamental contribution to the theory of excitable media.

Currently, the theory of active media relies mainly on reaction-diffusion type equations [NiP, H2, Mic, MiL, Mur, VRY, K9, K10, K11], such as

$$\frac{\partial X(R,t)}{\partial t} = F[X(R,t)] + \frac{\partial}{\partial R_i}\left(D_{ij}(X)\frac{\partial X}{\partial R_j}\right). \tag{42}$$

Here $X(R,t)$ is a set of macroscopic functions characterizing the system, for instance, the concentrations of chemical reactants; $F(X)$ are nonlinear functions determined by the structure of the system and the nature of the processes; and D_{ij} are the coefficients of spatial diffusion of the elements of the system. Here and bellow $r \equiv R$.

Specific examples of equations like (42) have been proposed and analyzed by Fisher (see [Mur]), Kolmomogorov, Petrovskii, Piskunov (1937), Zel'dovich, and Turing. To the same type also belong the various modifications of the Ginzburg-Landau equation, which is widely employed in the theory of equilibrium and nonequilibrium phase transitions.

Equations of reaction-diffusion type describe a broad class of physical, chemical, and biological phenomena. We begin with an example of the spatial diffusion of the independent van der Pol oscillators (generators).

If in the description of the generators, we restrict ourselves to information concerning the energy of oscillations $E(R,t)$, then the appropriate Kolmogorov-Petrovskii-Piskunov (KPP) equation for this function with the constant coefficient of space diffusion D has the form

$$\frac{\partial E(R,t)}{\partial t} = (a - bE(R,t))E(R,t) + D\frac{\partial^2 E}{\partial R^2}, \quad a = a_f - \gamma. \tag{43}$$

Here γ and b are linear and nonlinear friction coefficients, respectively, and a_f is the feedback parameter.

This equation, however, is not by itself sufficient to give a complete statistical description of the distributed system of generators, because, apart from the spatial diffusion, there is another cause that disturbs the dynamic regime of generation. This is due to the effects of noise on the internal degrees of freedom of the generators with the intensity D_E. The combined influence of both factors is taken into account in the equation for the distribution function $f(E,R,t)(\int dE\frac{dR}{V}f = 1)$ [K5, K6, K9]

$$\frac{\partial f}{\partial t} = \frac{\partial}{\partial E}\left(D_{(E)}E\frac{\partial f}{\partial E}\right) + \frac{\partial}{\partial E}[(-a + bE)Ef] + D\frac{\partial^2 f}{\partial R^2}. \tag{44}$$

Only in the special case in which noise acting on the internal degrees of freedom of the generator is negligibly small $(D_{(E)} \equiv 0)$, does equation (44) have a particular solution

$$f(E,R,t) = \delta(E - E(R,t)); \; <E> = \int_0^\infty EfdE = E(R,t) \tag{45}$$

that corresponds to the one-moment approximation. The function $E(R,t)$ satisfies the FKPP equation (43).

7 Kinetic and Hydrodynamic Description of the Heat Transfer in Active Medium [K10, K11]

Now we shall go back to the generalized kinetic equation (15) with the collision integrals (16) and (21). Let us suppose the mean is $u(r,t) = 0$ and, therefore, the distribution function is $f = f(r, |\ v\ |, t)$ (see (29)), from this kinetic equation, the heat transfer equation follows

$$\frac{\partial < E >}{\partial t} = \chi \frac{\partial^2 < E >}{\partial R^2}, < E >= \frac{3}{2} æ T(r,t). \tag{46}$$

From the kinetic equation follows the corresponding equation for the dispersion of temperature

$$\frac{\partial}{\partial t} < (\delta E)^2 >$$
$$= \frac{4}{\tau} \left(\frac{2}{3} < E >^2 - < (\delta E)^2 > \right) + \chi \frac{\partial^2 <(\delta E)^2>}{\partial R^2} + 2\chi \left(\frac{\partial <E>}{\partial R} \right)^2. \tag{47}$$

We see that the source of the temperature fluctuations is defined by the gradient of the mean temperature $T(r,t)$. Therefore, for the complete description of the heat transfer, it is necessary to use the kinetic equation for the distribution $f(R, |\ v\ |, t)$.

Let us now assume that a nonlinear source of the heat exists, then the constant $1/\tau$ is replaced by the nonlinear dissipative coefficient $\gamma(E)$. We shall suppose (for illustration only) that

$$\gamma(E) = -a_s + \frac{1}{\tau} + bE, \text{ and } D_{(v)}(E) = \left(\frac{1}{\tau} + bE \right) æ T_o. \tag{48}$$

Here the coefficient a_s characterizes the source of heat, and T_o is the thermostat's temperature. In the equilibrium state, the Maxwell distribution is with the temperature T_o.

Now we can represent the generalized kinetic equation (15) in the form

$$\frac{\partial f(R, |\ v\ |, t)}{\partial t} + v \frac{\partial f}{\partial R} = \frac{\partial}{\partial v} \left[D_{(v)}(E) \frac{\partial f}{\partial v} \right] + \frac{\partial}{\partial v}(\gamma(E)vf) + \chi \frac{\partial^2 f}{\partial R^2}. \tag{49}$$

In one-moment approximation, we have the nonlinear heat transfer equation

$$\frac{\partial < E >}{\partial t} = 2[\frac{3}{2}mD(< E >) - \gamma(< E >)] < E > + \chi \frac{\partial^2 < E >}{\partial R^2}. \tag{50}$$

Thus, in this approximation we obtain the reaction diffusion equation type (42). Here, however, not only is the space diffusion D, taken into account, but we also take into account the internal diffusion $D_{(v)}$. It is possible, therefore, to obtain solutions for all values of the bifurcation parameter a_s.

In the two-moment approximation from the kinetic equation (49), follows the system of corresponding reaction diffusion equations.

To illustrate the effectiveness of the kinetic approach to describe processes in an active medium, it is useful to consider the stationary solution. If the functions $\gamma(E)$ and $D(E)$ are defined by the expressions (48), then we can represent the stationary solution in the form

$$f(v) = C\exp\left[-\frac{mv^2}{2\mathit{æ}T_o} + \frac{a_s}{\mathit{æ}T_o b}\ln\left(1 + \frac{b}{\gamma}\frac{mv^2}{2}\right)\right]. \tag{51}$$

This describes the velocity distribution for all values of the rule parameter a_s. At $a_s = 0$ (51) coincides with the Maxwell distribution.

8 Kinetic Equation for Active Medium of Bistable Elements

We have considered some examples of the kinetic equations for active media, taking into account both the spatial diffusion and the diffusion with respect to the internal variables of the elements of the medium.

Let us suppose that the nonlinear force is defined by the expression

$$F(x) = -m\omega_o^2(1 - a_f + bx^2)x. \tag{52}$$

Here a_f and ω_o are the feedback parameter and eigenfrequency of the linear oscillator, respectively.

Let us also suppose that $f(x, v, R, t)$ is the distribution function of the internal variables x, v, and a space position R of the bistable element, then we can present the generalized Fokker-Planck equation in the following form:

$$\frac{\partial f}{\partial t} + v\frac{\partial f}{\partial x}\frac{F(x)}{m}\frac{\partial f}{\partial v} = \frac{\partial}{\partial v}\left(D_{(v)}\frac{\partial f}{\partial v}\right) + \frac{\partial}{\partial v}(\gamma v f) +$$
$$+ \frac{\partial}{\partial x}\left[D_{(x)}\frac{\partial f}{\partial x} - \frac{F(x)}{m\gamma}f\right] + D\frac{\partial^2 f}{\partial R^2}. \tag{53}$$

We see that there are two internal diffusion coefficients:

$$D_{(v)} = \gamma\frac{\mathit{æ}T}{m}, \qquad D_{(x)} = D_{(o)}(1 + bx^2) \quad (D_{(o)} = \frac{\mathit{æ}T}{m\gamma}), \tag{54}$$

and the diffusion coefficient D of elements in space R. For the equilibrium state, the Maxwell-Boltzmann distribution is the solution of the kinetic equation (53).

If we restrict the class of distribution functions (as in (29)) by condition,

$$f(x, v, t) = f(x, |v|, t), \tag{55}$$

then we can obtain by integration on v the corresponding generalized Einstein-Smoluchowsky equation for the function $f(x, R, t)$

$$\frac{\partial f}{\partial t} = \frac{\partial}{\partial x}\left[D_{(o)}(1 + bx^2)\frac{\partial f}{\partial x}\right] - \frac{\partial}{\partial x}[m\omega_0^2(1 - a_f + bx^2)xf] + D\frac{\partial^2 f}{\partial r^2}. \tag{56}$$

It is useful to remember that, in the theory of stochastic processes, the transition from the Fokker-Planck (the Kramers) equation to the Einstein-Smoluchowsky equation is made by using the perturbation theory on the small parameter $F'/m\gamma^2$ (or $\omega_0^2/\gamma^2 \ll 1$) [VK, Ris, Gar, K6]. This method corresponds to the Hilbert, Chapman-Enskog, and Grad perturbation theory for the Boltzmann equation on a small Knudsen number.

We see that in the stochastic theory it is possible to avoid using the perturbation theory on the correspondingly small "Knudsen parameter."

It is interesting to consider some limiting cases.

(1) The distribution function for the stationary homogeneous state

$$f_{St} = C \exp\left[-\frac{U_{eff}(x)}{æT}\right], \quad U_{eff} = \frac{m\omega_o^2}{2}\left[x^2 - \frac{a_f}{b}\ln(1 + bx^2)\right]. \qquad (57)$$

At $a_f = 0$, expression (57) coincides with the Boltzmann distribution for $U = m\omega_o^2/2$.

(2) The space distribution function $f(R,t)$. From (56), the self-diffusion equation follows

$$\frac{\partial f}{\partial t} = D\frac{\partial^2 f}{\partial R^2}, \quad \int f(R,t)\frac{dR}{V} = 1. \qquad (58)$$

(3) *The one-moment approximation for incompressible medium.* If $D_{(o)} \equiv 0$, then the distribution function has the form

$$f(x, R, t) = \delta(x - x(R,t)), \quad x(R,t) = <x> . \qquad (59)$$

For the "field function" $x(r,t)$ from the generalized Einstein-Smoluchowsky equation follows the FKPP-type equation

$$\frac{\partial x(R,t)}{\partial t} = \gamma\left[a_f - 1 + bx^2(R,t)\right]x(R,t) + D\frac{\partial^2 x(r,t)}{\partial R^2} \cdot \gamma = \frac{\omega_o^2}{\gamma}. \qquad (60)$$

In this approximation, the internal diffusion coefficient is $D_{(x)} \equiv 0$. The diffusion $D_{(x)}$ plays an important role.

Let us suppose that the space diffusion is one-dimensional, in the direction ζ, then there exists the following well-known stationary solution of equation (60)

$$x(\zeta) = \left(\frac{a_f - 1}{b}\right)^{1/2} \text{th}\left[\left(\frac{a_{f-1}}{2D}\right)^{1/2}\zeta\right]. \qquad (61)$$

The relative width of the front is determined by the expression $(2D/(a_f - 1))^{1/2}$. We see that this result is not valid in the region near the bifurcation point $a_f = 1$.

To obtain a more general solution for all values of the bifurcation parameter, it is necessary to use the kinetic equation [K9].

9 Kinetic Fluctuations in Active Media

9.1 The Langevin Source in the Kinetic Equation

The collision integrals are defined by the small-scale (kinetic) fluctuations. We shall now consider the large-scale (kinetic) fluctuations whose dynamics are determined by the kinetic equations themselves.

Here are two well-known methods of calculating the kinetic fluctuations [K1, K2, K3, K4, K5, K6, LiP, VK, GGK, KoS, Kei]. One method is based on the approximate solution of the set of equations for the moments of random (pulsating) distribution functions. The second method is based on solving the corresponding Langevin equations for the random distribution function $\tilde{f}(x, R, t)$, and this is what we will use here.

In the first method, for example, we introduce a Langevin source $y(r, x, t)$ into the kinetic equation (56) for the active medium of bistable elements. In the Gaussian approximation, the two moments of the Langevin source are given by the following expressions [K5, K9]:

$$< y(x, R, t) >= 0, \qquad < y(x, R, t)y(x', R', t') >=$$

$$2\left[D_{(x)}\frac{\partial^2}{\partial x \partial x'} + D\frac{\partial^2}{\partial R \partial R'}\right]\delta(x - x')\frac{1}{n}\delta(R - R')f(x, R, t)\delta(t - t'). \qquad (62)$$

The distribution function $f =< f >$ is defined by (56).

These general expressions give us the possibility of finding the Langevin sources in the diffusion equation for the space distribution $\tilde{f}(R, t) = \int \tilde{f}(x, R, t)dx$ and in the FKPP equation (60).

9.2 Spatial Diffusion. "Tails" in the Time Correlations

From the kinetic equation (56) with the Langevin source (62) follows the equation for the fluctuations of space distribution $\tilde{n} = n\tilde{f}(R, t)(n = N/V$, $n(R, t) =< \tilde{n}(R, t) >)$

$$\frac{\partial \delta n}{\partial t} = D\frac{\partial^2 \delta n}{\partial R^2} + y(R, t), \qquad < y(R, t) >= 0,$$

$$< y(R, t)y(R', t') >= 2D\frac{\partial^2 f}{\partial R \partial R'}\,\delta(R - R')n(R, t)\delta(t - t'). \qquad (63)$$

In this way, we have come to the well-known equation of spatial diffusion with the random source. The distribution $n(R, t)$ is the solution of the diffusion equation (58).

In the equilibrium state, the spectral density of fluctuation is defined by the well-known expression

$$(\delta n \delta n)_{\omega, k} = \frac{2Dk^2}{\omega^2 + (Dk^2)^2}n. \qquad (64)$$

Using the Fourier transformation over ω and k, we find the expression for the space-time correlation on the fluctuations δn

$$< \delta n \delta n >_{r, \tau} = \frac{n}{(4\pi D \mid \tau \mid)^{3/2}}\exp\left(-\frac{r^2}{4D \mid \tau \mid}\right), \qquad r = r - r', \qquad \tau = t - t'. \quad (65)$$

We see that at $r = 0$ the time correlations fall off according to the power law $\propto 1/|\tau|^{3/2}$.

It is necessary to emphasize that the results presented here were obtained without taking into account the boundary conditions. In other words, the results are valid only for an infinite medium. If the size of system L is finite, then these results are valid only at conditions

$$\omega \gg \tau_D^{-1}, \quad \tau \ll \tau_D = L^2/D. \tag{66}$$

The role of the boundaries will be dealt with in Section 10.

9.3 The Langevin Source in the Reaction Diffusion (FKPP) Equation

For the incompressible medium, the integral $\int f(x, R, t) = const$, therefore, the correlator of the source $y(R, t)$ in the diffusion equation – the correlator (63) now is zero, but the moment of the Langevin source $y_{(x)}(R, t) = \int xy(x, R, t)dx$ is not zero and is now defined by the expression

$$y_{(x)}(R, t) >= 0, \quad < y_{(x)}(R, t)y_{(x)}(R', t') >=$$
$$2\left[D_{(o)}(1 + b < x^2 >_{(R,t)}) + D\frac{\partial^2}{\partial R \partial R'} < x^2 >_{(R,t)}\right]\frac{1}{n}\delta(R - R')\delta(t - t'). \tag{67}$$

To obtain the closed expression for this correlator, it is necessary to use the solution of the generalized Einstein-Smoluchowsky equation (56). The situation is much simpler when the stationary solution can be used (57). In this case, $< x^2 >$ in (67) for all values of the bifurcation parameter a_f is defined by the following expression:

$$< x^2 >= C \int x^2 \exp\left[-\frac{m_o\omega_o^2}{2\varkappa T}\left(x^2 - \frac{a_f}{b}\ln(1 + bx^2)\right)\right]dx. \tag{68}$$

Thus, in the stationary state (68), the intensity of noise is defined by the mean-square value $< x^2 >$, therefore, for the calculation of the fluctuation $\delta x(R, t)$, it is possible to use the self-consistent approximation on $< x^2 >$ [K3, K4, K9]

$$\frac{\partial \delta x(R, t)}{\partial t} + \gamma_x \delta x(R, t) - D\frac{\partial \delta x(R, t)}{\delta R^2} = y(R, t). \tag{69}$$

With the help of this equation and the definition (67) for the intensity of noise, we obtain the following expression for the spectral density of fluctuations

$$(\delta x \delta x)_{\omega,k} = \frac{2(\gamma_x + Dk^2)}{\omega^2 + (\gamma_x + Dk^2)^2}\frac{< x^2 >}{n}. \tag{70}$$

We introduced here the special definition for the half-width of the spectral line, which is determined by the internal diffusion $D_{(o)}$ in any bistable element

$$\gamma_x = \frac{D_o}{< x^2 >}(1 + b < x^2 >). \tag{71}$$

This definition is valid for all values of the bifurcation parameter a_f.

It is necessary to state that the results presented in this section and the corresponding results in the previous section are valid only for an infinite medium – the condition in (66).

10   Natural Flicker Noise ("1/f Noise")

10.1 Natural Flicker Noise for Diffusion Processes [K4, K5, K12, K13, K14]

For a fluctuative diffusion process, we can distinguish two domains: $\tau_{cor} \ll \tau_D$ and $\tau_{cor} \gg \tau_D$. For the first region, the size of the system has little effect on the fluctuations. The spectrum in this case practically coincides with the spectrum (64).

Natural flicker noise exists in the low-frequency region

$$\frac{1}{\tau_{obs}} \ll \omega \ll \omega_{\max} = \frac{1}{\tau_D} = \frac{D}{L^2}. \tag{72}$$

Thus, the upper limit of the region is determined by the diffusion time. The minimum frequency is determined by the time of observation τ_{obs}, which is limited only by the lifetime of the device.

From (72), it follows that the actual volume of diffusion $V_\omega \equiv L_\omega^3 =$ depends upon the frequency and is much larger than the volume V of the sample,

$$V_\omega \equiv L_\omega^3 = (D/\omega)^{3/2} \gg V. \tag{73}$$

Under this condition, the size of the system is not important and, in the limit, the sample can be treated as a point (dimension zero).

To find the spectral density of the Langevin source, it is necessary to take into account that the diffusion volume V_ω for the region of flicker noise is V_ω/V times larger than the volume of the sample V. Accordingly, the mean concentration n is replaced by the effective concentration [K12]

$$n \to n_{eff} = AV_\omega < \delta n_V \delta n_V > . \tag{74}$$

Here we have used δn_V to denote the one-point correlator of fluctuations smoothed over the volume of the sample V. The constant A will be defined below. As a result, the expression for the spectral density of the Langevin source now takes the form

$$(y,y)_{\omega,k} \equiv 2Dk^2 n_{eff} \exp\left(-\frac{L_\omega^2 k^2}{2}\right). \tag{75}$$

The corresponding expression for the spectral density of δn is

$$(\delta n \delta n)_{\omega,k} = \frac{2Dk^2}{\omega^2 + (Dk^2)^2} AV_\omega < \delta n_V \delta n_V > \exp\left(-\frac{L_\omega^2 k^2}{2}\right). \tag{76}$$

From these expressions, the variance of the distribution of wavevector is $< (\delta k)^2 >= L_\omega^{-2} = \omega/2D$ and tends to zero as $\omega \to 0$. In this way, we come to the coherent distribution in the space of wave numbers.

After integrating over k, we find the expression for the temporal spectral density of natural flicker noise

$$(\delta n \delta n)_\omega = \frac{\pi}{\ln(\tau_{obs}/\tau_D)} \frac{< \delta n_V \delta n_V >}{\omega}, \qquad \tau_{obs}^{-1} \ll \omega \ll \tau_D^{-1}. \tag{77}$$

The constant A in (76) is defined here from the condition

$$\int_{1/\tau_{obs}}^{1/\tau_D} (\delta n \delta n)_\omega \frac{d\omega}{\pi} =< \delta n_V \delta n_V > . \tag{78}$$

Thus we assume that the main contribution to the integral over ω comes from the region of flicker noise. This expression can be written in the form

$$\frac{(\delta n \delta n)_\omega}{n^2} = \frac{2\pi a}{N \omega}, \qquad a = \frac{1}{2\ln(\tau_{obs}/\tau_D)} \frac{< \delta n_V \delta n_V >}{n/V}. \tag{79}$$

Here we have introduced the notation a for Hoog's constant [Kog]:

The time correlator for the region of natural flicker noise is given by the expression [K12]

$$< \delta n \delta n >_\tau = [C - \ln(\tau/\tau_D)/\ln(\tau_{obs}/\tau_D)] < \delta n_V \delta n_V >;$$
$$\tau_D \ll \tau \ll \tau_{obs}, \quad C = 1 - \gamma/\ln(\tau_{obs}/\tau_D), \quad \gamma = 0.577. \tag{80}$$

We see that the dependence on τ in the region of flicker noise is very weak (logarithmic with a large argument), therefore, we may speak of the residual time correlations.

The Langevin Equations for the Function δn in the Region of Flicker Noise. The Langevin equation for the Fourier component $\delta n(\omega, k)$ can be presented in the form

$$(-i\omega + Dk^2)\delta n(\omega, k) = y(\omega, k). \tag{81}$$

The spectral density of the source is given by (75).

Because the distribution over the wave numbers in the region of flicker noise is very narrow, in (81) we can replace

$$k^2 \to< k^2 >_k = L_\omega^{-2} = \omega/D \tag{82}$$

and use a simpler equation for the fluctuations $\delta n(\omega)$,

$$(-i\omega + \gamma_\omega)\delta n(\omega) = y(\omega); \qquad \gamma_\omega = |\omega|. \tag{83}$$

The expression for the spectral density of the Langevin source is

$$(yy)_\omega = 2\gamma_\omega \frac{\pi}{\ln(\tau_{obs}/\tau_D)} < \delta n_V \delta n_V >; \qquad \gamma_\omega = |\omega|. \tag{84}$$

Thus, for the region of natural flicker noise the dissipative coefficient $\gamma_\omega =\mid \omega \mid$, and tends to zero as $\omega \Rightarrow 0$. This expression relates the spectral density to the dissipative coefficient and one-time correlator.

In this way, we have formulated the principal results of the author's theory of flicker noise [K12, K13, K14, K4, K5]. The natural flicker noise arises whenever the final stage of relaxation toward the equilibrium state is associated with spatial diffusion.

The dependence on the actual structure of the system only enters via two parameters: the time of diffusion τ_D and the one-time correlator $< \delta n_V \delta n_V >$. Recall that we are considering the diffusion of a physical entity of any kind.

This theory has been employed elsewhere to explain the existence of natural flicker noise in music, as observed by Voos and Clarke [VoC]. We considered the possible connection between natural flicker noise and superconductivity [K14, K5]. So far, these two fundamental phenomena have been treated independently. We demonstrated, however, that the existence of low-frequency natural flicker noise in a system of charged Bose particles results in the vanishing of electrical resistance and thus facilitates the existence of a permanent superconducting current and the Meissner-Oxenfeld effect.

The transition from the normal to the superconducting state is a well-known example of a phase transition of the second kind, which results in the appearance of a macroscopic quantity of charged Bose particles (Cooper pairs). To understand the important role of natural flicker noise in a system of Cooper pairs for the existence of permanent superconductivity electrical current, it is necessary to take into account that the appearance of natural flicker noise, according to this theory, implies the creation of a coherent state in the space of wave numbers at $\omega \Rightarrow 0$. In this way, we are talking about the linkage between two coherent processes.

We can remark that the natural flicker noise does not exist only in a system of Cooper's pairs. In a system of neutral Bose particles (He^4), the existence of natural flicker noise makes superfluidity in narrow gaps possible. The coefficient of diffusion is then of the order of Planck's constant h.

Naturally, there also exists "technical" noise with a $1/f$ spectrum, which is associated, for example, with mobile defects.

It is very interesting also to remark that after integrating over ω (not over k!), from (76) we find the spatial natural flicker noise "$1/k$." Thus, not only does temporal natural flicker noise exist, but spatial natural flicker noise also exists – the noise "$1/\mid k \mid$."

10.2 Natural Flicker Noise for Reaction-Diffusion Processes

We saw that for statistical description of processes in active media it is more natural to use the corresponding generalized kinetic equations instead of the reaction-diffusion equation (42). We saw also that such an approach provides the possibility of describing kinetic fluctuations. One of the methods of describing kinetic fluctuations is based on solving the corresponding Langevin equations.

For example, in the Einstein-Smoluchowsky equation for active media of bistable elements, the moments of the Langevin source are defined by expressions (62).

We used this general expression to define the Langevin source in the generalized reaction-diffusion equation for the "field" variable $x(R,t)$. As a result, we obtained the expression (70) for the spectral density of fluctuations. This result is valid for all values of the bifurcation parameter a_f.

As for the diffusion processes (see (64)), this result was obtained without taking the boundary conditions into account, therefore, this method is valid only for an infinite medium for the region of frequencies $\omega \gg \tau_D^{-1} = D/L^2$.

For the diffusion systems, the natural flicker noise exists in the low-frequency region (72). For reaction-diffusion systems, not only is dissipation determined by diffusion, the additional internal dissipation is also determined. If the corresponding coefficient $\gamma_{(x)} \ll \tau_D^{-1}$ exists, there is then a low-frequency region of flicker noise

$$\gamma_{(x)} = \omega_{\min} \ll \omega \ll \omega_{\max} = 1/\tau_D = D/L^2. \tag{85}$$

To find the corresponding spectral function, it is necessary to change the second term (67) in correspondence with the definition (75). After integration over k, we again obtain the expression for the temporal spectral density for the low-frequency region (85)

$$(\delta x \delta x)_{\omega,k} = \frac{2(\gamma_{(x)}+ \mid \omega \mid)}{\omega^2 + (\gamma_{(x)}+ \mid \omega \mid)^2} A < \delta x_v \delta x_v > . \tag{86}$$

The constant A is defined by the expression

$$A = 4/\pi \int_{\gamma_{(x)}}^{1/\tau_D} \frac{2(\gamma_{(x)}+ \mid \omega \mid)}{\omega^2 + (\gamma_{(x)}+ \mid \omega \mid)^2} d\omega. \tag{87}$$

The dissipative coefficient $\gamma_{(x)} \equiv \omega_{\min}$ is defined by the expression (71) for all values of the bifurcation parameter a_f.

11 Criteria of Self-Organization

11.1 Evolution in the Space of Controlling Parameters. S-Theorem

Assume that we have made the choice of controlling (rule) parameters a, and shall consider the evolution of the sequence of stationary states corresponding to different values of the rule parameters.

We single out a state corresponding to $a = a_0$ and another state with $a = a_0 + \Delta a$. We take the state with the distribution function $f_0(x, a_0)$ as the state of physical chaos (the correctness of this assumption will have to be verified). We also introduce the distribution function $f(x, a_0 + \Delta a)$ and represent it as an effective "canonical Gibbs distribution"

$$f(x, a_0 + \Delta a) = \exp \frac{F - H(x, a_0 + \Delta a)}{D};$$

$$f_0 = f(x, a_0 + \Delta a)|_{\Delta a=0}. \tag{88}$$

The distribution functions f, f_0 are normalized in the same way

$$\int f \, dx = \int f_0 \, dx = 1. \tag{89}$$

To find the actual form of the distribution (88), we need a mathematical model of our system. However, because in many cases it is difficult to construct a mathematical model, in the next section we will give a modified formulation of the criterion of the self-organization, which allows us to define the effective Hamiltonian function directly from experimental data without using a mathematical model.

For accessing the relative orderedness of states with different values of a according to the entropy values, we must renormalize expression (88) to the given value of the energy [K15, K16, K17, EbK, K5, K6].

We shall formulate the S-Theorem in two steps.

First, as an additional condition, we fix the mean energy $< H(x, a_0) >$ for the state of physical chaos and renormalize $f_0 \Rightarrow \tilde{f}_0$. The function $\tilde{f}_0$ is presented as

$$\tilde{f}_0(x, a_0, \Delta a) = \exp \frac{\tilde{F}_0 - H(x, a_0)}{\tilde{D}(\Delta a)}, \qquad \int \tilde{f}_0 \, dx = 1. \tag{90}$$

The additional condition has the form

$$\int H(x, a_0) \tilde{f}_0(x, a_0, \Delta a) \, dx = \int H(x, a_0) f(x, a_0 + \Delta a) \, dx. \tag{91}$$

From the solution of this equation, we find the effective temperature $\tilde{D}$ as a function of Δa

$$\tilde{D} = \tilde{D}(\Delta a); \qquad \tilde{D}(\Delta a)|_{\Delta a=0} = D. \tag{92}$$

By $\tilde{S}_0$, we denote the entropy for the state with the distribution $\tilde{f}_0$. Then, subject to the conditions (89) and (91), the difference between $\tilde{S}_0$ and S is given by

$$\tilde{S}_0 - S = \int \ln \frac{f}{\tilde{f}_0} f \, dx \geq 0. \tag{93}$$

So we have obtained two results, (92) and (93). The change in the degree of order upon the transition $a_0 \Rightarrow a_0 + \Delta a$ is assessed from the solution (92). If the inequality

$$\tilde{D}(\Delta a) > D \tag{94}$$

is satisfied (that is, if the effective temperature in the state with $a = a_0$ must be increased to make (91) valid), then $a \Rightarrow a_0 + \Delta a$ is the transition from a less ordered state (physical chaos) to a more ordered state. The difference in the entropies (93) is a quantitative measure of the increase in the degree of order.

Also, if the inequality (94) is not satisfied, then the change in a is not a controlling one. This is an indication that we have to look for new controlling parameters. Thus, the S-Theorem serves as a tool for checking the correct choice of the controlling parameters.

If there are several controlling parameters, the search for the most ordered states can be optimized (see ref. in [K4, K5]). The above statement was termed the S-Theorem, with "S" standing for "self-organization," to emphasize that the S-Theorem is a criterion of self-organization.

11.2 The Comparison of the Relative Degree of Order of States on the Basis of the S-Theorem Using Experimental Data

Practical applications of the criterion for the relative degree of order based on the S-Theorem, as outlined above, require a knowledge of the structure of the effective Hamiltonian function. It is very important to have criteria that can use the experimental data directly.

Such criteria can be based on K-entropy, Lyapunov indices, and fractal dimensions, which can be obtained from the realizations of the processes concerned. Let us show that the relative degree of order can be accessed directly from experimental data using the S-Theorem criterion [K17].

As above, we start by choosing the controlling parameter a. The state with $a = a_0$ is taken as that of physical chaos, with which the state for $a = a_0 + \Delta a$ will be compared.

We use the experimental realizations $x(t, a_0), x(t, a_0 + \Delta a)$ as the chosen set of internal parameters x of our process. The realizations must be sufficiently long, so that they can be used to obtain the distribution functions

$$f_0(x, a_0), \quad f(x, a_0 + \Delta a); \quad \int f_0\, dx = \int f\, dx = 1. \qquad (95)$$

Using the distribution f_0, which by our assumption corresponds to the state of physical chaos, we find the function

$$H_{eff} = -\ln f_0 \qquad (f_0 = \exp(-H_{eff})), \qquad (96)$$

which in the renormalized distribution $\tilde{f}_0$, which will serve as the effective Hamiltonian function. From (95) and (96) it follows that we need no extra information to find H_{eff}, except the known realization $x(t, a_0)$. The mean value of the effective energy for the distributions (95) in general will depend on Δa. Let us now renormalize to the preset value of $< H_{eff} >$. We introduce the renormalized distribution $\tilde{f}_0$, which we represent as the canonical distribution

$$\tilde{f}_0(x, a_0, \Delta a) = \exp \frac{F(D) - H_{eff}(x, a_0)}{D(\Delta a)}, \quad \int \tilde{f}_0\, dx = 1. \qquad (97)$$

The effective free energy F as a function of the temperature D is found from the normalization condition of the function $\tilde{f}_0$. The effective temperature D as a function of Δa is derived by solving the equation

$$\int H_{eff}(x, a_0) \tilde{f}_0(x, a_0, \Delta a)\, dx = \int H_{eff} f(x, a_0 + \Delta a)\, dx, \qquad (98)$$

that is, by requiring that the mean effective Hamiltonian function H_{eff} be constant. Using the solution of this equation, we find the required function

$$D(\Delta a); \quad D(\Delta a)|_{\Delta a=0} = 1, \qquad \Delta a \geq 0. \qquad (99)$$

Now we again use the distributions $\tilde{f}_0$, f to find the difference between the entropies

$$\tilde{S}_0 - S = \int \ln \frac{f}{\tilde{f}_0} f \, dx \geq 0 \qquad at \qquad < H_{eff}(x, a_0) >= const. \qquad (100)$$

If the solution of (99) is such that

$$D(\Delta a) > 1; \qquad D(\Delta a)|_{\Delta a=0} = 1, \qquad (101)$$

then the state with $a = a_0 + \Delta a$ is more ordered than the state with $a = a_0$, which we took for the state of physical chaos. This conclusion must, however, be verified (see [K17, K5]).

It is useful now to remark on the following:

During the time evolution to the equilibrium state, we were dealing with degradation processes.

Dealing with the evolution of stationary states in the space of controlling parameters, we encounter a new possibility. The fact is that the controlling parameters can be found among the parameters that characterize the stationary state. Changes in the latter may lead to the reduction of the entropy, thus resulting in self-organization. In this way, the formulation of the second law of thermodynamics is extended: The evolution of stationary states in the space of controlling parameters may be associated both with increasing and decreasing entropy. In the latter case, the set of parameters that define the stationary state includes the controlling parameters.

These two types of evolution represent, in a sense, the two extremes. It would be interesting to investigate a more general case in which the parameters change during the time evolution.

12 Conclusion. Associative Memory and Pattern Recognition

Since the well-known work by Hopfield (1982), the problems of associative memory and pattern recognition occupy an important place in statistical physics, in particular, in the theory of spin glasses in optical systems.

The models considered in the theory of associative memory and pattern recognition are mostly based on discrete neural networks, each neuron possessing a number of discrete states. Haken (1988) proposed treating the recognition of patterns as a process similar to the formation of dissipative structures in synergetic systems. In the proposed models, the dynamics of the system are determined by a potential function that depends on the parameters. This dependence may be used for storing the information concerning the pattern to be recognized. The test patterns are introduced via the initial components of the vector of the dynamic system. Systems of this kind exhibit the phenomenon of associative memory, because they are capable of restoring the complete pattern in the process of evolution toward the stationary state, starting with the incomplete information contained in the initial conditions [H3, FuH].

Let us demonstrate that the simplest medium capable of associative memory can be constructed from active elements, which are van der Pol generators interacting via common feedback. Such a system is practical and feasible.

Consider the following set of equations [K9]:

$$\frac{dX_i}{dt} + \frac{1}{2}\sum_{j=1}^{N}(-a_j + b_j E_j)x_j = V_i, \quad i = 1, 2, \ldots, N,$$

$$\frac{dV_i}{dt} + \frac{1}{2}\sum_{j=1}^{N}(-a_j + b_j E_j)V_j + \omega_i^2 x_i = 0, \qquad E_i = \frac{1}{2}(V_i^2 + \omega_i^2 X_i^2). \quad (102)$$

The generators are linked together via the common feedback. In general, all the coefficients a_i, b_i are different. Observe that the excitation of all generators in the system requires only that X_{0i}, V_{0i} be nonzero for at least one generator.

Introducing the "dissipative potential" of a system of generators

$$U = \sum_i U_i, \qquad U_i = \frac{b_i}{2}(E_i - \frac{a_i}{b_i})^2, \qquad (103)$$

we may write these equations in the form

$$\frac{dX_i}{dt} + \frac{1}{2}\sum_j \frac{\partial U}{\partial X_j}\frac{1}{\omega_j^2} = V_i; \qquad \frac{dV_i}{dt} + \frac{1}{2}\sum_j \frac{\partial U}{\partial V_j} + \omega_i^2 X_i = 0. \qquad (104)$$

In the stationary state, the dissipative potential (103) goes to zero. The stationary solution with fixed initial phases ϕ_{0i} depends on N parameters a_i/b_i. This dependence can be used for storing information about a pattern of N points. For restoring the pattern, it is sufficient to introduce into the initial parameters the information about at least one point of the pattern. Then, in the process of evolution toward the stationary state, the system will restore complete information about the pattern installed into the stationary state via the values of the parameters a_i/b_i. Obviously, this model is just a very simple illustration. Possible generalizations may proceed in various directions. Of interest, for instance, is the statistical generalization, when the state of the system in the general case is characterized by "N-particles" distribution function f_N. We then enter the domain of the statistical theory of "nonideal" active systems with interaction of different "particles," which can simulate the processes of associative memory and pattern recognition with due account for fluctuative processes, involving both the internal variables of the system's macroscopic elements and the variables that describe the motion of each element as a whole [K5, K6].

References

[K1]   Klimontovich, Yu.L.: On Nonequilibrium Fluctuations in a Gas. TMF **8** 1971, 109.

[K2]   Klimontovich, Yu.L.: Kinetic Theory of Non Ideal Gases and Non Ideal Plasmas. Nauka, Moscow, 1975; Pergamon Press, Oxford, 1982.

[K3]   Klimontovich, Yu.L.: The Kinetic Theory of Electromagnetic Processes. Nauka, Moscow, 1980; Springer, Berlin, Heidelberg, 1983.

[K4]   Klimontovich, Yu.L.: Statistical Physics. Nauka, Moscow, 1982; Harwood Academic Publishers, New York, 1986.
[K5]   Klimontovich, Yu.L.: Turbulent Motion and the Structure of Chaos. Nauka, Moscow, 1990; Kluwer Acad. Pub., Dordrecht, 1991.
[K6]   Klimontovich, Yu.L.: Statistical Theory of Open Systems. Kluwer Academic Publishers, Dordrecht, 1994.
[MY]   Monin, A.S., Yaglom, A.M.: Statistical Fluid Mechanics. Nauka, Moscow, 1965; MIT, 1971.
[LL]   Landau, L.D., Lifshitz, E.M.: Fluid Mechanics. Nauka, Moscow, 1986; Pergamon Press, Oxford, 1959.
[K7]   Klimontovich, Yu.L.: Statistical Theory for Non Equilibrium Processes in a Plasma. Nauka, Moscow, 1964; Pergamon Press, Oxford, 1967.
[Kr]   Krylov, N.S.: Works for the Foundation of Statistical Physics. Moscow, Nauka, 1950.
[Pr]   Prigogine, I.: From Being to Becoming. Freeman, San Francisco, 1980; Nauka, Moscow, 1985.
[PrS] Prigogine, I., Stengers, I.: Order out of Chaos. Heinemann, London, 1984; Progress, Moscow, 1986.
[RSC] Romanovski, Yu.M., Stepanova, N.V., Chernavsky, D.S.: Mathematical Biology. Nauka, Moscow, 1984.
[K8]   Klimontovich, Yu.L.: Entropy Evolution in Self-Organization Processes. H-Theorem and S-Theorem. Physica A **142** 1987, 390.
[Lor] Lorenz, E.: Deterministic Nonperiodic Flow. J. Atm. Sci. **20** 1963, 167.
[H1]   Haken, H.: Synergetics. Springer, Berlin, Heidelberg, 1978; Mir, Moscow, 1980.
[Ani] Anishchenko, V.S.: Complex Oscillations in Complex Systems. Nauka, Moscow, 1990.
[NeL] Neimark, Yu.I., Landa, P.S.: Stochastic and Chaotic Oscillations. Nauka, Moscow, 1987; Kluwer Acad. Publ., Dordrecht, 1992.
[LiP] Lifshitz, E.M., Pitaevsky, L.P.: Statistical Physics. Nauka, Moscow, 1978.
[NiP] Nicolis, G., Prigogine, I.: Self Organization in Non Equilibrium Systems. Wiley, New York, 1977; Mir, Moscow, 1979.
[H2]   Haken, H.: Advanced Synergetics. Springer, Berlin, Heidelberg, 1983; Mir, Moscow, 1985.
[Mic] Michailov, A.S.: Foundations of Synergetics I. Springer, Berlin, Heidelberg, 1990.
[MiL] Michailov, A.S., Loskutov, A.Yu.: Foundations of Synergetics II. Springer, Berlin, Heidelberg, 1991.
[Mur] Murray, G.: Lectures on Nonlinear Differential Equations Models in Biology. Clarendon Press, Oxford, 1977.
[VRY] Vasiliev, V.A., Romanovsky, Yu.M., Yachno, V.G.: Autuwaves. Nauka, Moscow, 1987.
[K9]   Klimontovich, Yu.L.: Some Problems of the Statistical Description of Hydrodynamic Motion and Autowaves Processes. Physica A **179** 1991, 471.
[K10] Klimontovich, Yu.L.: On the Necessity and the Possibility of the Unified Description of Kinetic and Hydrodynamic Processes. TMF **92** 1992, 312.
[K11] Klimontovich, Yu.L.: The Unified Description of Kinetic and Hydrodynamic Processes in Gases and Plasmas. Phys. Let. A **170** 1992, 434.
[VK]   Van Kampen, N.G.: Stochastic Processes in Physics and Chemistry. North-Holland, Amsterdam, 1983.
[Ris] Risken, H.: The Fokker-Planck Equation. Springer, Berlin, 1984.

[Gar]   Gardiner, C.W.: Handbook of Stochastic Methods for Physics, Chemistry, and
         Natural Sciences. Springer, Berlin, Heidelberg, 1984.
[GGK]   Gantsevich, S.V., Gurevich, V.L., Katilus, R.: Theory of Fluctuations in Non
         Equilibrium Electron Gas. Rivista del Nuovo Cimento **2** 1979, 1.
[KoS]   Kogan, Sh.M., Shul'man, A.Ya.: To the Theory of Fluctuations in a Nonequi-
         librium Gas. ZhETF **56** 1969, 862.
[Kei]   Keizer, J.: Statistical Thermodynamics of Nonequilibrium Processes. Springer,
         Berlin, Heidelberg, New York, 1987.
[K12]   Klimontovich, Yu.L.: Natural Flicker Noise. Pis'ma v ZhTF **9** 1983, 406.
[K13]   Klimontovich, Yu.L., Boon, J.P.: Natural Flicker Noise (1/f-noise) in Music.
         Europhys. Lett. 3 **4** 1987, 395.
[K14]   Klimontovich, Yu.L.: Natural Flicker Noise (1/f-noise) and Superconductivity.
         Pis'ma v ZhETF 51 **1** 1990, 43.
[Kog]   Kogan, Sh.M.: The Low Frequency Current Noise with Spectrum 1/f in Solid
         State, Usp. Fiz. Nauk, **145** 1985, 285; Sov. Phys. Usp. **28** 1985, 171.
[VoC]   Voos, R.F., Clarke, J.: "1/f Noise" in Music: Music from 1/f Noise. J. Acoust.
         Soc. Am. 643 **1** 1978, 258.
[K15]   Klimontovich, Yu.L.: Entropy Decrease in the Processes of Self-Organization.
         S-Theorem. Pis'ma v ZhTF **9** 1983, 1089.
[K16]   Klimontovich, Yu.L.: S-Theorem. Z. Phys. B. **66** 1987, 125.
[K17]   Klimontovich, Yu.L.: Problems in the Statistical Theorie of Open Systems:
         Criteria for Relative Degree of Order of States in Self-Organization Processes.
         Usp. Fiz., Nauk **158**, May 1989, 59; Sov. Phys. Usp. 32 **5** May 1989.
[EbK]   Ebeling, W., Klimontovich, Yu.L.: Selforganization and Turbulence in Liquids.
         Teubner, Leipzig 1984.
[H3]   Haken, H.: Synergetic Computers and Cognition. A Top-Down Approach to
         Neural Nets. Springer-Verlag, Berlin, Heidelberg, 1991.
[FuH]   Fuchs, A., Haken H.: In Neural and Synergetic Computers. Ed. H. Haken,
         Springer-Verlag, Berlin, 1988, 16.

Index

Printing: Mercedesdruck, Berlin
Binding: Buchbinderei Lüderitz & Bauer, Berlin